SMITHSONIAN CONTRIBUTIONS TO KNOWLEDGE.

202

GEOLOGICAL RESEARCHES

IN

CHINA, MONGOLIA, AND JAPAN,

DURING THE YEARS 1862 TO 1865.

BY

RAPHAEL PUMPELLY.

[ACCEPTED FOR PUBLICATION, JANUARY, 1866.]

THIS memoir, having been approved by the National Academy of Sciences, has been accepted for publication by the Smithsonian Institution.

JOSEPH HENRY,
Secretary S. I.

COLLINS, PRINTER,
PHILADELPHIA.

PREFACE.

THE material for the following pages was collected since 1860. Leaving the Eastern States in that year, and crossing the plains to Arizona, I remained there nearly a year in charge of silver mines. Being forced by the Indian troubles to abandon that territory, I entered Mexico, and after a midsummer journey over the deserts of the Pacific coast, between Sonora and California, reached the latter State.

Leaving California with one companion, Prof. William P. Blake, both of us engaged by the Japanese Government to explore the island of Yesso, we sailed for Japan *via* the Sandwich islands. The engagement with the Japanese Government lasted but little more than a year, when it was suddenly brought to an end by the fierce political troubles of that time. It was during hasty journeys of reconnoissance that the notes relating to Yesso were jotted down, and at a time when I hoped to be able to make a much more thorough study of the geology of Japan.

It was with true regret that I left the service of a government whose courtesy had made a lasting impression on my memory, and with whose struggles for progress as against exclusiveness I deeply sympathized.

Crossing to China, after a short visit to Nagasaki, I ascended the Yangtse Kiang into Central Hunan, and to the frontier of Sz'chuen, a great part of the journey being made in a small Chinese boat, and occupying four months of the spring and summer of 1863.

The autumn and winter of 1863 and spring of 1864 were spent in examining the Coal fields west of Peking, for the Chinese Government, and in journeys in Northern China and Southern Mongolia.

I spent the summer of 1864 at Nagasaki.

In the winter of 1864 and 1865, in company with Mr. T. Walsh, of Japan, and Mr. F. R. St. John, Secretary of the British Legation at Peking, I crossed into Siberia, and thence, alone, travelled overland to St. Petersburg and Paris.

Thus the journeys which furnished the data for the following pages were as follows:—

I. In 1862 over the ground indicated in the sketch map of southern Yesso, Pl. No. 8, and excursions in the neighborhood of Yokohama.

II. In 1863 excursions in the vicinity of Nagasaki; a journey up the Yangtse Kiang to the boundary between Hupeh and Sz'chuen, and into southern Hunan; and excursions from Peking into the mountains of northwestern Chihli.

III. In 1864 a journey in southern Mongolia, along the edge of the plateau to near the great N. E. bend of the Hwang Ho, returning to Peking by a route south of the plateau and within the Great Wall; and finally, part of the journey homeward, from China across the plateau and the Gobi desert to Siberia.

With the exception of the itinerary in Yesso, which was made while in the service of the Japanese Government, and the description of the coal basin west of Peking, which was examined at the request of the Chinese Government, all the material was collected on journeys made at my expense.

Ignorance of the Chinese and Mongolian languages, the difficulty of making observations in western China, owing to the hostility of the people at the time, the intense cold of the winter journey across the plateau into Siberia, and the fact that the enterprise was a private one, will, it is hoped, serve as excuses for asking the indulgence of the reader in view of the incompleteness of the work.

I have attempted throughout to keep the generalizations separate from the record of observations and other data on which they rest.

I have followed, generally, the orthography of Dr. S. W. Williams for Chinese proper names, and that of Klaproth for Mongolian names, where these could be found on his great map of Central Asia, but in many instances they are written from the pronunciation of the Tartar guides. In giving Japanese and Aino names I have followed very closely the Japanese spelling.

For assistance in preparing the present work I am indebted to Dr. J. S. Newberry for undertaking the description of the fossil plants, and to Mr. Arthur Mead Edwards for the examination of infusorial earths, etc., under the microscope, and to Prof. G. J. Brush and Mr. James A. Macdonald for analyses of coals.

A considerable amount of valuable material consisting mainly of Paleozoic, Tertiary, and Post-tertiary shells, and of rocks, has not yet been worked up.

I would return thanks to Prof. J. D. Whitney both for many valuable hints, and for the use of his excellent library.

I am deeply indebted to Dr. W. Lockhart, Mr. C. Murray, and Dr. S. W. Williams, and Rev. Mr. Edkins, of Peking, for valuable assistance in making researches in Chinese geographical literature.

The diagrams in the text, and the plates, I. to VIII., at the end, are executed in copper relief engraving by Messrs. E. R. Jewett & Co. of Buffalo; plate IX. is cut in wood by Mr. C. Murry, of New York.

R. P.

NEW YORK, Aug. 1, 1866.

CONTENTS.

LIST OF DIAGRAMS.

LIST OF PLATES.

GEOLOGICAL RESEARCHES

IN

CHINA, MONGOLIA, AND JAPAN.

CHAPTER I.

ON THE GENERAL OUTLINES OF EASTERN ASIA.

If we examine a Mercator Chart of Eastern Asia, we are instantly struck with the parallelism of many of its most important features. A straight line (*A*, *B*, Pl. VII) drawn in the longer axis of the Gulf of Pechele, trending nearly northeast (N. 47° E.), if prolonged in both directions, will be found to coincide with the entire middle course of the Yangtse, between Sz'chuen and Yunnan, with the longer axis of the great delta-plain between the highlands of Shantung and western Chihli, with the mouth and lower course of the Liau river, with the valley of the lower Amur, and finally crossing the Sea of Ochotsk, it is parallel to, and nearly coincides with, the direction of the Gulf of Penjinsk.

Using this line as a standard of reference, we find that the long straight western shores of the two greatest indentations, the Sea of Ochotsk and the Bay of Bengal, are nearly in a line with each other and parallel to our standard. The same may be said of a line connecting the islands of Formosa, Kiusiu, Nippon and the Kuriles. The trend of the southeastern coast of China, the upper course of the Yellow river, the Lake Baikal, and the courses of many of the principal rivers of Eastern Siberia; that of Kamtschatka and the coast of Manchuria are all separate instances confirming this rule.

We are naturally led to look for the cause of this in a similar uniformity in the trend of the mountain ranges, and, indeed, although the directions of these are difficult of determination, I hope to be able to show that such a parallelism really exists. The long, submerged chain represented by the Kurile and Japanese islands is an unmistakable instance, while, in the northern part of the continent, the Stanovoi and Yablonoi ranges, and all the ridges of Trans-Baikal, are examples of mountains nearly or quite parallel to our standard, and inclosing extensive longitudinal valleys. The same may be said of the Byrranga mountains, and of almost all the ridges east of the Lena river. Indeed, while the trends of nearly all the mountains of North-eastern Asia lie between N. N. E. and E. N. E., the majority of them approach very nearly the N. E. S. W. direction.

Having seen that this regularity exists in the ranges of the better explored parts of Eastern Asia, let us look for it in China also, where we have to rely on a more limited number of data, partly geological and partly topographical in their character.

Where the Yangtse river crosses the Sz'chuen-Hupeh frontier, it cuts through a broad mountain range whose principal axis crosses the river in long. 111° 15′, near Ichang (fu). Here the axial granite rises 600 to 1000 feet above the river, and is flanked on both sides by an immense thickness of limestone and coal-bearing rocks, whose strata have here a mean trend to N. E. If, through this point, we draw a line (*C*, *D*, Pl. VII) having a similar trend, its prolongation will indicate the watershed between the Hwai river and the Han river, the watershed of Shantung, and following the line of islands that stretch across the entrance to the Gulf of Pechele, it will coincide with the range of mountains, which, beginning with the promontory of Liautung, divides the waters first of the Liau river and Yaluh river, and afterwards, of the Sungari river and Usuri river. If we prolong the line from the Yangtse to the S. W., it will nearly coincide with the mountains that part the rivers of Kweichau from those of Hunan.

All these ridges I take to be members of a continuous line of elevation, extending from Southern China to the Amur river, and which, from its influence on the character of the country, may be called the central anticlinal axis of China.

A line drawn from near Canton and passing through the Chusan archipelago, will represent the mean trend of the coast range, and, if prolonged to the N. E., it will cut the Corean peninsula near its southern end, in what appears to be its most mountainous point.[1] In the other direction, the island of Hainan, from its N. E. S. W. trend and lofty mountains, would seem to be a member of the same range.

In Northwestern China, a great range crosses the Yellow river, in its course between Shansi and Shensi, and trending N. E. by E., connects the mountain knot of Northwestern Sz'chuen with that of the Ourang daban north of the Tushïkau gate of the Great Wall. Nearly parallel to this is another range which, beginning west of Singan (fu), crosses the Yellow river, forming the Lungmun gorge, and traversing, obliquely, the centre of Shansi, gradually approaches the other range in northern Chihli.

These are the three principal axes, and they seem to be made up of parallel anticlinal ridges. Minor parallel axes seem to occupy the country between these larger ranges.

If we examine the maps of the provinces that border on the eastern edge of the Tibetan highland, we find a system of ranges, which, branching off from the Kwenlun and following, at first, a southeasterly course, gradually merge into a N. S. trend. The easternmost of these, occupying western Sz'chuen, divide the principal northern tributaries of the Yangtse. Those farther west form the narrow watersheds between the upper courses of the Yangtse, the Cambodia and the Salween, and, in their southern prolongation, they form the Malayan peninsula and probably that occupied by Annam and Siam. The N. S. trend seems to be confined exclusively to the extreme west of China.

---

[1] According to the great map of Kanghi this peninsula seems to have its principal mountains in the south, forming a N. E. S. W. ridge.

On the other hand the E. W. system of trends, which is so important in Central Asia, exercises an influence which is apparent much farther eastward.[1]

A range of mountains, said to have several snow-covered peaks, originating in Southern Kansuh, runs due east, separating the waters that enter the Yellow river through the Wei and the Loh, from those that flow to the Yangtse through the Kialing and the Han, and finally disappears in western Honan. Another range, with a mean E. by S. trend, is given by Klaproth as forming the boundary between Sz'chuen on the south and Shensi and Kansuh on the north.

It is not improbable, that the country included between these two ranges in Shensi and Kansuh, is an elevated table-land. The courses of the Han and Kialing rivers and the communication between their waters, as indicated by Chinese authorities, seem to favor this idea.

In the south, the Nanling mountains, a range said to have peaks that reach above the snow-line, rise in Yunnan, and, branching, form, in the northern member, the boundary between Kwangsi and Kweichau, while the southern member trends off into Kwangsi. The influence of the northern branch of the Nanling, is apparent as far as Fuhkien, in the probably comparatively low watershed north of Kwangtung. The higher portion of this range seems to be along the southern boundary of Kweichau, where it has lofty peaks and fertile elevated table-lands,[2] which, from difficulty of access, have been for ages the home of the aboriginal Miautsz, a race unconquered by the surrounding civilization. The two passes that cross this range in Hunan and Kiangsi, where it is called the Meiling, cannot be very high, as the portage between the head of boat navigation on the two flanks is only a few miles. According to Biot,[3] the members of Lord Amherst's embassy give the height of the Kiangsi pass as 3000 feet. The great map of Kanghi gives an uninterrupted water communication between the headwaters of the Siang river of Hunan and those of a tributary of the Si river, that flows through the city of Kweilin.

I have here attempted to trace only those ridges which seem to be the most important, as exhibiting the general configuration of China. To the E. W. ranges is due the fact, that the mean courses of the great rivers of the empire lie east and west. But the total length of each river is made up of N. E. reaches, where it flows through broad and fertile longitudinal valleys, and of southeasterly or southerly reaches in which it traverses, by deep and narrow gorges, the N. E. S. W. ridges.

---

[1] All that is known of these two systems, the N. S. and the E. W. is derived from the Jesuit maps and from Chinese writers.

[2] Chinese Repository, I. 40.

[3] Recherches sur la hauteur, etc., Journ. Asiat., 1840.

CHAPTER II.

GEOLOGICAL OBSERVATIONS IN THE BASIN OF THE YANGTSE KIANG.

A GLANCE at the section (Pl. 1) across Central China will show that the Devonian limestone and Chinese Coal measures seem to predominate, at least at the surface, over all else. There is only one point in the whole length of the section, where rocks older than the great limestone deposit rise to the surface, so that if the former exist, they are buried deep below the level of the sea. I shall give, in a subsequent chapter, reasons for believing that, at least in the valley of the Yangtse, there are also no representatives of the Mesozoic formations of later date than the Chinese Coal measures, and few, if any, of the Cenozoic.

Where the Yangtse breaks through the ridges of the central anticlinal axis of elevation, in Eastern Sz'chuen and Western Hupeh, a section, nearly eighty miles long, is exposed in the succession of deep gorges through which the river passes this barrier. Here the Devonian limestone is seen to rest almost immediately on the granite, a comparatively small development of metamorphic schists intervening.

This seems to be the only point between Western Sz'chuen and the Pacific, where the Yangtse has exposed these lower rocks, and even here they occur during only about eight miles of the river's course, and with a maximum height of only a few hundred feet above the river. To their occurrence are due the rapids that render the navigation of this part of the "Great River" so dangerous.

The granite immediately above the first rapids consists of a triclinic feldspar and orthoclase, the former predominating, a brilliant black mica and quartz with small crystals of sphene scattered through the mass. Above Shantowpien the granite becomes very fine-grained, and still further up the river it is succeeded by syenitic granite, composed of white triclinic feldspar, quartz, large laminæ of brown mica, and crystals of hornblende, with minute octahedrons of magnetic iron.

On its eastern and western declivities the granite supports the metamorphic strata. Those to the eastward, which could not be closely examined, seemed to be gneiss trending E. W. and dipping about 30° to S. West of the granite the strata consist, where examined, of hornblendic schist and chloritic schist, the former often containing lenticular masses and cross veins of quartz, feldspar, and chlorite. Rolled fragments of diorite, probably of metamorphic origin, indicate the presence of this usual companion of these rocks. Near their contact with the granite these strata trend N. N. E., dipping about 85° to E. S. E., while further up the river their trend changes to E. N. E., and the dip to N. N. W.

Flanking this granite core on both sides and covering it, is the great Devonian limestone floor of the Chinese Coal measures. On the eastern flank of the granitic axis the limestone strata trend, almost uniformly, N. E. S. W., varying in dip from 25° to 8° towards the S. E. as we recede from the granite. On the western flank the strike is less regular, changing from nearly N. S., at the contact with the metamorphic schists, to N. E. S. W. in the upper part of the limestone. In the immediate neighborhood of the river, over an area of forty or fifty square miles, the limestone has disappeared, but in the distance, on both sides of the Yangtse, its yellow cliffs are seen towering to a height of more than 2,000 feet above the water.

I know of no limestone deposit that can rival this in thickness. Taking the length of the cross section from its contact with the younger conglomerates, near Ichang, to where it rests on the metamorphic schists, to be seven and one-half geographic miles, and the mean dip at 15°, viz., 10° for the eastern half and 20° for the western, we obtain the enormous thickness of 11,600 feet, more than two statute miles. I observed no faults in this gorge, and the great thickness observed in this same limestone in Northern China, leads me to think that the above estimate cannot be far from the truth.

West of this ridge of limestone is another of about the same size, the intervening space being occupied by the Coal measures.

Here, within a distance of eighty miles, are the principal rapids, while the river traverses the limestone through a series of five gorges unsurpassed in the grandeur of their scenery. The Yangtse, which, a few miles below the mouth of the Ichang gorge, has a width of 960 yards, is in this narrowed to 250, and in the Fungsiang gorge to 150 yards.[1] In these narrow passages, whose walls are from 900 to 1200 feet high, cliffs of bare rock, often vertical or overhanging, alternate with steep declivities clothed in green from the water to the summit, and with deep, inaccessible dells filled with the rich growth of a semi-tropical vegetation. Streams flowing from the mouths of caverns high above the river, cool the air in their descent, while the huge clusters of stalactite which they have formed—the work of ages—show well the chemical power of the smallest drop, side by side with the mechanical force of the rolling river. Through these gloomy chasms the skilful boatmen drag the heavy junks, now "tracking" them from paths and steps hewn in the solid rock, now pulling them by rusty and time-worn chains clamped along the vertical walls.

The depth of the water must be very great,[2] and the difference between high and low water is said to be as much as eighty feet in the Ichang gorge.

The limestone is generally of a bluish-gray color and compact texture, though subordinate to this variety, layers occur having every shade of color and grain. A gray, compact variety, with frequent large crystals of calcite is not uncommon; and a very compact, almost black kind is quarried in the Ichang gorge. Indeed gray, pink, red, black, and blue varieties of this same limestone, with compact, porphyritic and crystalline textures, furnish in almost every province of China

---

[1] Blackiston. Five months on the Upper Yangtse.

[2] Blackiston's party found no bottom with eighteen fathoms.

useful and choice marbles. Every degree of thickness occurs in the layers from laminæ only one-quarter inch thick to beds of many feet.

Nodules and thin layers of black chert occur throughout the limestone, but in the lower half they are remarkably frequent, becoming more common as we approach the oldest beds, in which, indeed, the calcareous rock is often entirely excluded by massive layers of quartzite. At the eastern entrance to the Lucan gorge, where the limestone rests on the older rocks, the lowest beds of the former, containing lenticular masses and thin layers of chert, are soon succeeded by a bed 40 to 50 feet thick, of massive quartzite.

Wherever I have had occasion to examine this limestone in place, it has invariably appeared to be entirely without fossils, but this has been only in the main ridges, where metamorphic action has probably played a more important part than in the minor ridges that rise between these lines of greater elevation, and it seems to me that there can be little doubt that the fossil Brochiopoda that occur in many provinces belong to this formation.

Just before entering the eastern mouth of the Lucan gorge, a bed of fine-grained, micaceous, gray sandstone is observable, intervening between the metamorphic schists and the limestone. The trend of this intervening bed is N. N. W. and the dip 25° to 30° to W. S. W., the metamorphic schists striking to E. N. E. and dipping to N. N. W., while the trend of the overlying limestone strata, at the nearest point observed, was about N. by W. and the inclination about 30° to W. by S.

At the western end of the Mitan gorge we enter the coal field of Kwei. Here the limestone disappears under strata, apparently conformable with it, of a fine-grained micaceous sandstone, which, below Kwei, is succeeded by a fine-grained, gray, calcareous sandstone. The trend of the beds which, near the gorge, was N. N. E. with a dip of about 40° to W. N. W., changes here to N. with a dip to E., and further up, opposite Kwei, it is N. by W. with an inclination of 70° to E. by N. Here is the beginning of a series of those angular plications so common to Coal measures in all countries. Small beds of limestone and red argillite alternate with the sandstones until, about two miles above Kwei, the first coal seams crop out, and with the appearance of these, the trend changes to N. W. by W., more than 90° from its normal direction of N. E. S. W.

The seams of coal are of an inferior friable anthracite. Those I visited above Kwei were highly inclined between sandstone walls, and contained, according to the Chinamen, only six to eight inches of fuel. Capt. Blackiston, who took specimens of these rocks and noticed, with much accuracy, the general features of this region, remarks that the rocks of the coal regions of Sz'chuen, wherever he saw them, presented the same appearance as those of the Kwei field.[1] It would seem probable that in Sz'chuen, which seems to be occupied by an immense coal basin, the Coal measures exist with a much greater thickness than in the Kwei field, where only the lower members seem to have been preserved. Deposits of iron ore occur in intimate connection with coal and limestone in Sz'chuen,[2] and, as we shall

---

1 Five Months on the Upper Yangtse.

2 Ibid.

see later, it is probable that the extensive salt deposits of that province are members of the same formation.

Near the city of Ichang, at the eastern mouth of the gorge, the limestone strata, trending here N. E. and dipping about 8° to S. E., are covered by apparently conformable beds of fine-grained, gray sandstone, which, toward the top, soon merges into a coarse conglomerate. The change is very marked, the upper portion of the sandstone containing rounded fragments of chert near the contact, and the lower part of the conglomerate having lenticular deposits of the sandstone. This transition appears to mark some important change that took place during the forming of these deposits, and the fact that, in transverse section, they border the river for twelve miles and have a great thickness, would seem to indicate that this change was not confined to the immediate neighborhood.

This conglomerate is followed by a red sandstone, which above Itu dips easterly, and below that place westerly. From here eastward the country on both sides of the river is flat, the rocks being covered for the most part by alluvial deposits; but in the neighborhood of Yangchi limestone crops out in different places, with a very irregular strike between N. and W., and a corresponding dip to between N. and E. From this point to Hankau, the country, if we except a few isolated hills, is one almost unbroken plain, the ancient bed of the Tungting lake, in which the older rocks are covered by the lake deposits.

At the town of Shishan (Hien) an isolated hill rises from the plain, its almost vertical strata trending about N. 65° E., and consisting of sandstone, arenaceous shale resembling a similar rock of the Kwei coal field, and a shaly quartzose conglomerate. The outcroppings of the older rocks that appear, at intervals, between the outlet of the Tungting lake and Hankau are sandstones and argillites, which, from their general character and the fact that in one place their trend is toward a locality a few miles distant where coal is worked, would seem to belong to the Coal measures. The hills immediately above Hankau are of clay slates and argillaceous sandstone, and through the cities of Wuchang and Hanyang, stretches a ridge of sandstone altered to an almost compact quartzite.

The journey from Hankau to the sea was made in a steamer, stopping only at Kiukiang and Chinkiang, making the knowledge concerning this part of the river very imperfect. The only sources of information were constant observations, through a good glass, of the frequent natural sections made by the river, and the scanty remarks of a few travellers connected with Lord Amherst's embassy.

Below Sankiangkau beds of sandstone and conglomerate, trending S. W. and dipping 40°—45° to S. E., are exposed, and a few miles further down the river the city of Hwangchau fu is built on a low ridge of ferruginous sandstone, of which the raised beds strike due N., dipping about 30° W. About twenty miles S. E. from this city, hills of limestone, 800 to 900 feet high, form the southern bank of the river, the irregular trend of their strata varying from W. to S. W., and the dip, of about 40°, from S. to S. E. Twenty-five miles below this point the river breaks through another ridge of limestone, the strata of which have a strike to S. E. by S. and incline about 40° to S. W. by W.

The rocks on the outlet to the Poyang lake have all the appearance of limestone, and this is the case with all the exposed sections from the outlet to the Siauku shan or Little Orphan rock. Below Tungliu coarse red sandstone is exposed, its upturned edges, which are here capped with the younger terrace deposits, trending to N. E. with a dip of 15° to N. W. At Nanking there are extensive quarries of limestone, while directly opposite the city, on the left bank of the Yangtse, strata of red sandstone trend W. S. W., dipping about 40° to E. S. E. Coal mines are worked in the immediate neighborhood of this city, especially on its eastern side. Soon after leaving the hills of Nanking the river enters the great delta plain through which it winds to the sea.

In a *résumé* I shall try, by means of a combination of the data given above, with information derived chiefly from native sources, to throw more light on the structure of this region.

TERRACES OF THE YANGTSE VALLEY.

At frequently recurring points along both the Upper and Lower Yangtse, we meet with deposits of gravel and clay, forming bluffs at the water's edge, or fringing the hills that form the walls of the valley. They are generally stratified in horizontal beds. Differing in height and in the character of their ingredients, there seems also to be a diversity of age. The extensive plain, once occupied by the Tungting lake, before it was reduced to its present size, is fringed by these terraces; for they recur constantly from Hankau to Yochau on the right bank of the river, and from this city along the eastern border of the lake, and form a belt which extends many miles to the south, and occupies nearly all the space along the southern edge of the lake, between the Siang and Yuen rivers. Again, where the river enters the lake plain, the tongue of land included by the river bend between Pahyang and Tung'sz, consists of the same deposit.

At the last named locality the deposit is made up of rounded pebbles of quartz and limestone, cemented with a stiff clay, and this is its general character at the junction of the Siang river with the lake and along the eastern shore. But the most general form of occurrence is that of a stiff blue clay, with irregular white spots. Near Tung'sz the terraces appear to be from seventy to ninety feet high, but below the outlet of the lake they vary from thirty to sixty feet. Blackiston mentions similar terraces as occurring at various points along the Yangtse in Sz'chuen.

The village of Tsingtan, at the eastern end of the Mitan gorge in Western Hupeh, is built on a terrace of conglomerate-breccia formed of fragments of limestone, chert, gneiss, and other metamorphic rocks, in form of rubble and rounded and angular fragments of all sizes, the whole firmly cemented by a calcareous tufa. This formation originally filled the valley from side to side, and its bluffs rise forty to fifty feet above high-water mark. In the rapid current that must always have scoured these narrow portions of the Yangtse valley, nothing but the coarsest material could resist the onward movement; and when an increase in the velocity of the stream took place, only those portions of the deposits were preserved which were near enough to the limestone to be cemented into a hard mass by the waters flowing from it.

The bed of the Yangtse must have been cut to about its present depth, when a diminution of its average fall took place, permitting the formation of these terrace deposits. Subsequently another change, by increasing the fall, caused the river to scour out, again, the greater part of the valley. As with the river so with the Tungting lake; this large sheet of water, which then occupied all the plain of Hupeh and Hunan, must have been filled up with the terrace deposit, the remains of which now form its shores. With the returning increase of fall, the lake was scoured out by the rivers Yangtse, Han, Siang, and Yuen. Since this erosion, it would seem probable that the velocity of the current has slightly diminished, as the material brought down by these rivers has converted nearly nine-tenths of the former lake into dry land. A large part of this lake-plain is said, by ancient Chinese writers, to have been an immense marsh where it is now cultivated land.

We have, at present, no observations to show whether the oscillations of Central China, which are thus recorded in the Yangtse Valley, were contemporaneous with the raising of the western edge of the delta-plain; but whether they were or not, the cause which was exerted across the whole breadth of China, must be looked for in a vertical movement, either in the Tibetan highland or along the eastern coast.

A remarkable instance of the formation of a deposit of fine material, in the swiftest part of the river, is observable in the first rapids, just above the Ichang gorge. Granite rocks rising to the surface, near the shore, form an obstruction to the current, which is here from fifteen to eighteen miles an hour, causing eddies in their lee, in which a constant precipitation of sand takes place. Banks of quicksands are thus formed, their tops almost even with the surface of the river. Their sides, too steep to remain at rest, are constantly being washed away, and as constantly replaced by the freshly precipitated material. At low water these banks line the shores, and, during the high water season of 1863, I noticed one more than half a mile long, and twenty-five or thirty feet above the river; the result of some previous very high freshet.

CHAPTER III.

OBSERVATIONS IN THE PROVINCE OF CHIHLI.

Along the western boundary of the province of Chihli, the great delta-plain is bounded by the outliers of the northwestern belt of N. E. S. W. ridges. The foundation on which rest the limestone and volcanic rocks of Northern Chihli, Shansi, and Shensi, consists of granite and the metamorphic schists; and where this foundation forms the northwestern limit of the delta-plain, it forms also the southeastern edge of the skeleton of the great table-land of Central Asia.

We have seen that, in Central China, the granitic and metamorphic rocks that support the limestone and Coal measures, rise to the level of the river, in, to say the least, only rare instances, and then as the axial cores of ridges; the great thickness of the overlying rocks making it highly probable that, from western Sz'chuen to the Pacific, this foundation lies far below the level of the sea. But if we cross the mountains from the delta-plain to the highlands of Mongolia, we find that the surface of the granitic substructure lies everywhere above the sea, and probably nowhere at a less height than 1000 feet. Were the limestone and younger rocks removed, the country would present the appearance of a table-land ribbed with high N. E. S. W. ridges, and very similar to southern Mongolia if we suppose that divested of its lava beds.

Along the edge of the plain, the limestone floor of the Coal measures rises abruptly from under the delta-deposit, and forms, so to speak, the eastern facing of these mountains. At the entrance to the Nankau pass, the strata trend N. 60° E. and dip about 40° to S. E. Five or six miles farther west, it is followed by granite, and between these points, strike and dip are very irregular. From the pass, the limestone stretches away to N. E. toward Jehol, and to S. W., facing the plain, toward Shansi.

While the Coal measures probably remain intact under the delta-plain, from the mountains of Shantung to those of Chihli, they exist in these latter only in scattered basins, where they have been partially preserved, by folds of the limestone, from denudation. The most important instances of this kind facing the plain, are the basins of Wangping (hien) and Fangshan (hien) west of Peking, and of Pingting (chau) in Shansi.

The basins of Wangping (hien) and Fangshan (hien) lie in the mountains west of Peking, where, rising from under the plain, they occupy synclinal folds of the limestone, and are probably only two arms of a larger basin concealed under the younger deposits to the eastward. The Wangping basin extends due west more than thirty miles, with a breadth of about twelve miles. Along a great part of its northern edge, a bed of porphyry conglomerate, of great thickness, intervenes between the limestone and the coal rocks, while the western portion of the basin is much broken up by porphyries, and the centre is crossed by a high ridge apparently of quartzose conglomerate and sandstones.

Coal seams, varying in thickness and quality, occur in many parts of these basins, and are worked in the more accessible localities, as, for instance, at Muntakau, Maanshan, the hill of Piyünsz, Lingchi on the Wangping creek and at Chaitang in the west.

In the following necessarily incomplete table, I have attempted to show the structure of those parts of these basins that came under my observation:—

6	Coal or anthracite alternating with beds of argillaceous shales, sandstones, gray quartzose conglomerate-breccias and compact red and green argillites.	Hsingshun and Tatsau.	Chaitang district.
5	Alternating beds of coal, argillaceous shales, and sandstones. Coal (Futau seam). Black under-clay. Micaceous quartzose sandstone. Quartzose conglomerate. Yellow argillaceous shale with impressions of plants. Outcroppings concealed for several hundred feet by terrace loam. Compact green argillite. Coarse gray sandstone and conglomerate. Compact argillite, mottled green and red. Coarse gray sandstone. Friable and argillaceous gray sandstone. Red calcareous clay slate. Greenish sandstone (with specks of chlorite). Red calcareous clay slate. Gray sandstone. Red calcareous clay slate Gray sandstone. Green quartzose conglomerate.		
4	Anagenite (quartz, feldspar, and mica sandstone). Argillaceous shales and compact sandstones alternating with seams of anthracite.	Muntakau.	
4	Ferruginous sandstone altered to quartzite. Quartzose conglomerate. Anthracite. Micaceous, and black argillaceous shales. Calcareo-argillaceous shale.	Maanshan.	
3	Anthracite. Micaceous, and black argillaceous shale. Calcareo-argillaceous shale.	Fangshan at Yingwo mine.	
3	Clay-slates (green, black and red). Greenish sandstone passing into greenish quartzose conglomerate.. Argillaceous shale.	Niuchauling.	
2	Conglomerate of porphyry, limestone and quartz. Porphyry conglomerate.	Hun Ho and Chaitang.	
1	Upper limestone. Black clay slate. Lower limestone (cherty).	Upper Yangtse and Province of Chihli.	

The porphyry conglomerates, No. 2, which, in places along the northern edge of the basin, have a thickness of not less than 2000 feet, are wanting in the eastern part. The parts of the series marked No. 3, form the oldest beds, and they rest immediately on the limestone in their respective localities. Between Nos. 3 and 4 the character and extent of the intervening beds were not observed. The connection between Nos. 4 and 5 is made on lithological grounds, the same green sandstone and green quartzose conglomerate occurring above the coal seams of Muntakau, and low down in the series at Chaitang.

Limestone.—Here, as on the Yangtse, a great development of limestone forms the floor of the Coal measures. Although no good opportunity occurred, in this region, for estimating its thickness, this is undoubtedly several thousand feet. It is generally divided into two nearly equal parts by a bed of clay slates; though independently of this, the upper and lower strata are characterized, the latter by an abundance of chert, and the former by comparative freedom from that mineral.

The limestone is generally compact and blue, but in places it is white and saccharoid; and black, pink, and dark red varieties occur. The chert is black, and is abundant in the lower half, occurring in nodules, and in layers varying in thickness from less than one line to over forty feet, beds of this size generally forming the bottom of the limestone. In the basin of Siuenhwa (fu), near the Great Wall, the limestone is highly siliceous, but almost always retains a white appearance.

This formation furnishes, here, as in almost every province of the empire, besides lime, the marble so much used in Chinese ornamental architecture, for bridges, tombstones, gateways, and the lions that guard the portals of all official buildings. The white saccharoid variety is very beautiful, but disintegrates so rapidly that, even in the dry climate of Peking, inscriptions on exposed monuments two hundred years old are barely legible.[1] The black variety, which is very compact, breaking with a conchoidal fracture, retains a perfectly fresh surface after centuries of exposure.

A quarry at the Maanshan has supplied lime for the capital during many centuries; the continued excavation having widened and deepened the valley, removing small hills and leaving, over an area of perhaps one square mile, a deposit that might well perplex an observer, were the cause not still at work. Almost every point in this area seems to have been the site of a lime-kiln, which has left its cone of concentric layers, consisting of half burnt limestone, chert, fragments of coal and ashes. As new kilns were built over and between old ones, the result is a bed, the ingredients of which have become cemented to a hard concrete, by the refuse lime. In this deposit, the stream of the valley has cut its channel, in places, forty to fifty feet deep, with vertical walls, without reaching the limestone bottom.

Caves are abundant in this limestone, and many of them are said to be of great extent. One which I visited, near Fangshan (hien), consists of a series of large

---

[1] There is a white variety, used in monuments near Peking, in which inscriptions of the Kin dynasty are perfectly fresh, as, for instance, that used in the grand marble arch of Kiyungkwan in the Nankau pass.

chambers extending nearly in a straight line. The first two of these only were visible, the entrance to the third having been closed by an imperial order, owing to a party of visitors having lost their way and perished.

These chambers are connected by passages, so small that they can be entered only by creeping on hands and knees. Their longest axis is at right angles to the strike of the strata, and forms a considerable angle with the dip. The floor is covered with stalagmite, which, in the centre of one chamber, seems to be at least forty feet thick, and is connected with the roof by immense columns of stalactite. Like many large caverns in China, this one is sacred to Buddha, of which deity there is a well executed high-relief sculptured in the wall of the entrance; and the small passages have been worn and polished by the knees of pilgrims during centuries.

I looked in vain at the face of the rock at the entrance, for some signs of a crack corresponding to the plane of these chambers.

Some of the deep and narrow ravines of the surrounding hills, seem to have been formed by the caving in of similar caverns.

In parts of the empire, these caves abound in fossil bones, which are excavated and used in medicine, under the name of "dragon's bones," "dragon's claws," etc.

This limestone, forming, as it does, the floor of the Coal measures, appears, surrounding the different basins of these, in highly inclined beds, forming as it were a narrow frame, or, having a gentler dip, it occupies a broader space.

Porphyry Conglomerate.—In the mountains that border the Wangping basin on the north and west, there are extensive masses and dykes of porphyry, which have raised and cut through the limestone in all directions. From the detritus of this intrusive rock, the beds of the lower Coal measures at Chaitang, which are equivalent to those marked No. 3 in the table, seem to have been formed. The reason for supposing this, is, that as we approach the northern edge of the Chaitang basin, we find the porphyry conglomerate underlying, in the form of a flat boss, the beds forming the lower half of No. 5 which are eminently characterized by two peculiar rocks, that marked as "compact green argillite" and the still lower ones, "green quartzose conglomerate." Further on we find, that the porphyry conglomerate contains interstratified beds of sandstone. The fragments that form this extensive member of the Chaitang series, are, for the most part, derived from the masses of porphyry nearest at hand. Thus near Chingtai they are chiefly green felsitic porphyry, similar to that forming dykes in the limestone at Hiamaling, a few miles distant, while, along the Hun river, red and green varieties predominate, intrusive masses of both kinds occurring in the neighborhood.

Fragments of limestone and quartz are frequent in the porphyry conglomerate, and would seem to characterize its upper portion. Thus I have indicated in the table two distinct varieties, though perhaps on insufficient grounds.

This conglomerate furnishes an important page in the history of the Coal measures in this region. It shows us that there had been an elevation of the limestone, perhaps caused or accompanied by the intrusion of the porphyries, before the overlying rocks were deposited. The presence of fragments of limestone, quartz, and porphyry, shows that these older rocks had been subjected to an extensive denudation.

In the narrow gorge, through which the creek finds its way from the Chaitang valley to the Hun river, the contact between the limestone and porphyry conglomerate is visible (Fig. 1). The limestone strata are cut through at a right angle, and are seen to dip about 80° to the S.

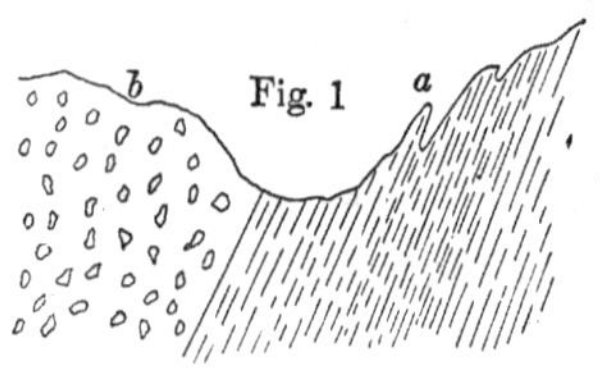

a. Upper limestone.
b. Lower porphyry conglomerate.

I did not obtain an observation of the dip of the conglomerate in this section to know whether it conforms to that of the limestone.

The coal district of Chaitang forms an area of low hills, and is limited on the north by the porphyry conglomerates, whose high and rugged hills are overtopped in the background by the yellow cliffs of the limestone. To the south rises a high ridge consisting, apparently, of the rocks of the Coal measures and dykes of porphyry, and separating the coal district of Chaitang from that of the Wangping creek. To the west is a high and hilly country mainly of porphyry.

About four miles W. N. W. of Chaitang, in the midst of this porphyry, lies the small coal district of Chingshui, and about five miles S. W. are the anthracite mines of the Tatsau district.

The valley of Chaitang has been occupied by a lake, the alluvial deposits of which now form terraces and cap hills over one hundred feet high. The trend of the tilted strata in the centre of the district is very uniformly N. W., and the dip is to N. E. and to S. W., forming both synclinal and anticlinal ridges. But as we approach the western end the trend becomes irregular, though the dip is toward the porphyry. Indeed, the edge of these mountains of porphyry, seems to mark the line of a great fault, perhaps combined with an immense overflow of that rock.

The following description of the more important coals is extracted from my Report to the Chinese Government, which is published in the "United States Diplomatic Correspondence, 1864, Part III."

For more perfect analyses of some of these and other coals by Mr. J. A. Macdonald, the reader is referred to Appendix No. 2.

Principal Mines.—The Futau mine, which lies about five li (less than two miles) S. S. E. of Chaitang, and from one hundred and fifty to two hundred feet above the level of the creek at that town, is remarkable as producing a "steam coal" that is equal if not superior to the best Welsh variety.

The seam, in which several openings have been made, is irregular in thickness, this varying from six to twelve feet, though in the mean averaging, probably, not less than seven feet. Near the roof the coal has a tendency to crumble, near the floor it is slaty; all the rest of the seam furnishes large blocks of firm and excellent fuel.

The coal has a brilliant lustre, is made up of well-defined layers, and has a tendency to a cubical fracture. It ignites quickly, burning with a long flame and little smoke.

Opening slightly, it burns without caking and without falling to pieces, and leaving a very little gray ash.

I found by dry assay, using the exceedingly imperfect means at my command in Peking, the following results:[1]—

Sp. gr.	1.31
Parts of lead reduced from oxide by one part of coal . .	31.50
Corresponding value in units of heat	7245.00[2]
Percentage of ash	4.00

There are several seams parallel to this one both above and below it, one of which is six or seven feet thick, and only thirty feet above it. The dip of the beds is about 45°.

So defective is the Chinese system of mining, that the proprietor of this mine could not undertake to furnish from it more than eight hundred and fifty tons yearly. The selling price, at the mouth of the mine, is $2 00 per ton of 2,000 pounds.

In the Fushun mine, apparently on the same seam, the coal reaches a thickness of thirty-five feet, though it averages much less.

Hsingshun Mine.—This is on one of a series of seams, that crop out in a valley about five li N. W. of Chaitang, and which I take to be younger than that of the Futau. The horizon of these seams is well characterized, in the Chaitang district, by the occurrence among them of beds of a peculiar quartzose conglomerate breccia, called by the natives horsetooth stone (from the appearance of pieces of chert it contains). This rock forms the floor of the seam in which lies the Hsingshun mine, while the roof is sandstone, and between these the seam dips at first 50°, changing gradually to 90°. Within a limited space the thickness of the coal varies from three to eight feet.

The coal is without lustre, and has an irregular flaky structure. It ignites quickly, burning with a long flame, cakes readily and leaves a red ash.

Sp. gr.	1.28
Parts of lead reduced by one part of coal	31.40
Units of heat	7222.00
Percentage of ash	3.00

The miners burn it in small heaps to a very light and porous coke.

Tatsau Mine.—About five miles S. W. of Chaitang is the Tatsau, or "great seam" of anthracite. It consists of two seams separated by about eight feet of sandstone, the upper one being from twenty-three to thirty-five feet thick, and the lower from seven to eighteen feet. The roof is formed by the same peculiar conglomerate breccia that characterizes the Hsingshun beds, the floor being sandstone, and dipping about 45° to N. W.

About six-tenths of the produce is anthracite of a superior quality, coming out in

---

[1] See Appendix No. 2.

[2] Without the correction of + $\frac{1}{8}$.

large, firm pieces formed of well-defined layers, with conchoidal fracture and brilliant metallic lustre.[1]

Sp. gr.	1.55
Parts of lead reduced by one part of coal	33.40
Units of heat	7682.00
Percentage of ash (gray)	4.00

Eight men produce about four tons daily, and the selling price at the mouth of the mine is $1 70 per ton. A short distance N.W. of the Tatsau is a high cliff of porphyry, forming part of the edge of the porphyry hills that bound the Chaitang district on the west. This rock is said, by the Tatsau miners, to cut off the coal and its accompanying rocks.

The annexed wood-cuts (Figs. 2 and 3) serve to give some idea of the Tatsau mine.

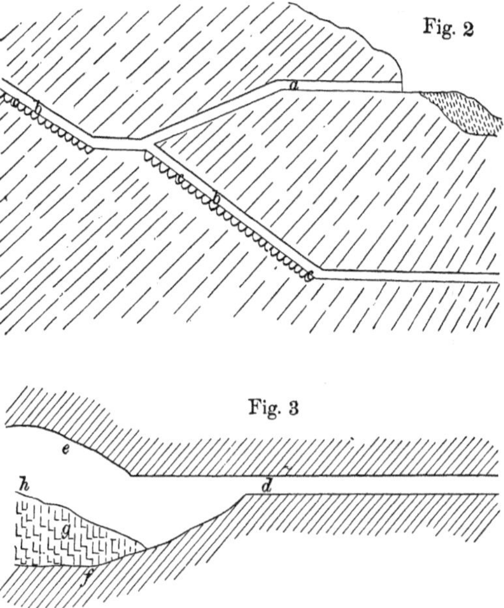

Fig. 2

Fig. 3

The entrance is by the gallery *a*, at first horizontal, then rapidly descending to the inclined shaft *b*. These are in the smaller and lower seam A drift leads to the level *d*, Fig. 3, in the larger seam. In working the coal the miners drive a level, as far below the surface as the amount of water will permit, and extending horizontally along the foot wall as far as the limits of the mine, with a breadth equal one-half of the seam when this is less than twenty feet. Beginning at the end *h*, they excavate the coal below the gallery, at *f*, to a depth of from ten to twenty feet. When this has advanced a short distance they break down from the

[1] See Appendix No. 2.

top *e*, and working back the coal is won from above and below the gallery at the same time, the refuse small coal, here about four-tenths of the whole, serving as a support *g*, in place of that extracted. The water is carried out by the inclined shaft *b*, fig. 2, the work being done by blind men, one of these standing in each of the hollowed out steps *c*, and bailing the water from his step to the one above him.

The coal is drawn out on sleds, by men, through *b* and *a*, only one-half the breadth of *b* being cut into steps for drainage.

Chingshui Mines.—These mines are in a narrow valley, about five miles W. N. W. of Chaitang, in the midst of the porphyry mountains. There seem to be several seams, but the confusion caused by the numerous dykes of porphyry is very great. In two of the seams the roof is formed by these dykes, at least for a considerable distance, while others are cut through by them, and in places only fragmentary portions of a seam, and its accompanying beds are left. Fig. 4 gives a general idea of the relation between some of the seams, and the porphyry as seen in the side of a mountain valley. Fig. 5 is a section of a fragment of the coal series only a few square

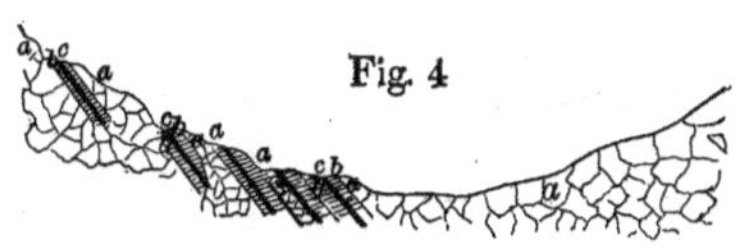

a. Porphyry. *b.* Coal series. *c.* Coal seams.

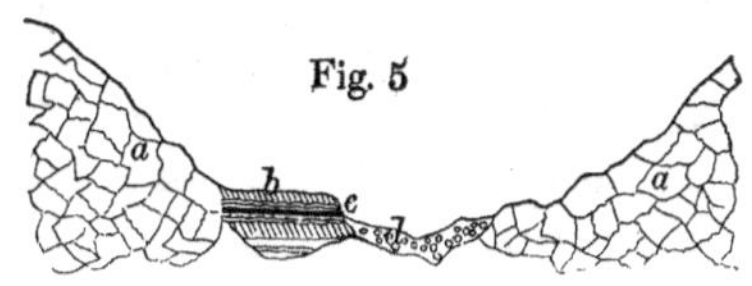

a. Porphyry. *b.* Coal series. *c.* Coal seam. *d.* Creek rubble.

rods in extent, cut off on one side by the porphyry, and on the other by the creek. The coal of this locality is very bituminous, and I failed, during my short visit, to find any indications of the metamorphism, often observed in the action of dykes on coal, especially where basalt has broken through tertiary brown coal formations.

The coal of the second seam from the right, Fig. 4 *c*,[1] is very brilliant, clean, and firm, breaking with a cubical fracture. It is very inflammable and melts and cakes, burning with a long flame, and leaving considerable ash.

Spec. gr.	1.38
Parts of lead reduced by one part of coal	29.00
Units of heat	6670.00
Percentage of ash	12.00

The seam from which this coal was taken had been worked about 500 feet on an incline, until stopped by water, and averaged between 7 and 8 feet in thickness. The fuel was best in the middle of the seam, and improved with the increasing depth. The proprietor worked two shifts of thirty men each, viz., eight miners, six carriers, ten water raisers, four men at mouth of mine, and two overseers. One miner produced, per shift, 1500 catties (about 1900 lbs.), of which two-thirds was coarse coal, and one-third fine.

[1] See Appendix No. 2.

The fuel, from this place, is almost all used in the tile-glazing establishments of Peking.

Porphyries.—In the mountains north of the Wangping coal basin, the limestone has been much disturbed by the intrusion of porphyry, which, in some places, traverses it in the form of large dykes, and in others rising under it in large dome-like masses, causes the overlying strata to dip from these in all directions.

As the porphyry conglomerates, at the bottom of the Coal series, are mostly derived from these rocks, their eruption took place before the Coal measures were deposited. Two varieties of felsitic porphyry were observed here, both younger than the limestone, and both represented in the conglomerate. One of these forms dykes on the ridge of Hiamaling and along the Hun Ho, between this ridge and Chingpaikau. At the first-named place, it incloses immense fragments of the black clay slate that divides the upper and lower members of the limestone.

This porphyry contains, in a compact, slightly greenish base, a little green mica and numerous crystals of a triclinic, milky-white and slightly opalescent feldspar, and is free from visible quartz. The feldspar weathers yellowish-red, and the base dirty-white. The rock strikes fire with the steel, though not very readily.

Near Yenchi, on the Hun Ho, a few miles below Hiamaling, is the second variety. It contains, in a light-pink base, crystals of feldspar, apparently orthoclase, and no visible quartz. The porphyry that cuts off the coal rocks near the Tatsau, is probably younger than the Coal measures, although it is uncertain whether it occurs in that locality as a dyke, or whether it is brought into the position it there occupies by a great fault.

This rock has, in a compact gray base, tending to green, numerous prisms of hornblende and small crystals of white feldspar, some of which at least are triclinic. It contains no visible quartz, and strikes fire with difficulty. Thus its characteristics are those of a hornblendic porphyry.

At Chingshui, two varieties of porphyry were observed, both traversing the coal rocks. In one of these, the base is black and fine-grained, containing numerous minute and small crystals of a transparent, colorless feldspar, certainly for the most part triclinic. There is no visible quartz, and the rock strikes fire with difficulty.

About ten miles S. E. of the entrance to the Nankau pass, near the granite point that juts out into the plain at Yangfang, there is an extensive fault in the limestone, the strata of this rock dipping toward the fault. Between the line of this fault and the granite there is a broad dyke of quartziferous porphyry. In a fine-grained pink base, it contains crystals of pink orthoclase and abundant grains of quartz.

It may not be out of place to mention here the coal districts of Muntakau and Fangshan. The former of these forms part of the Wangping basin where this disappears under the plain of Peking. The valley of Muntakau forms in itself a small bay, containing terraces of the plain deposit; there are said to be thirteen seams of anthracite in the sides of the valley, most of which have been worked since during the Ming dynasty.

Those seams which I visited alternate with sandstones and argillaceous shales, and underlie the peculiar green quartzose conglomerate that characterizes the lower part of the Chaitang series.

The Tehyih mine seems to be the most important, and has been worked for a horizontal distance of 8,500 feet. The seam is very irregular in thickness, varying from a mere thread to six or seven feet, and as much so in strike and dip. The anthracite is dull and hard and made up of layers. It flies to pieces in burning.[1]

Spec. gr.	1.79
Parts of lead reduced by one part of coal	31.00
Units of heat	7130.00
Percentage of ash	7.00

In this mine one miner produces on an average only about 100 catties—133 lbs.—daily, and the loss of time in bringing the coal to the surface is very great, the man who drags the sled being obliged, from the lowness of the gallery, to go on his knees the entire distance of more than a mile and a half. The men protect their knees and hands with cushions, a precaution of which I was able to appreciate the value after having gone in about 5,000 feet and back without any such protection.

The galleries grow smaller as the mine grows older, for, in replacing the old timber it often happens that the miners dare not remove an old piece, but are obliged to place the new one under it, and in this way the lapse of time reduces the height of the only thoroughfare of the mine. I was surprised on seeing at the entrance a very large fan-blower, made much like the machines used for fanning rice (which, in turn, are the same as our own fanning machines), and which is used here for ventilation.

In the district of Fangshan all the coal is said to be anthracite. Several seams are traversed by the galleries of the Yingwo mine, the lowest seam being only about 150 feet above the limestone, the intervening beds consisting of argillaceous shales, and the whole apparently conformably stratified with the limestone. The strike of these beds is E. W., and the dip about 30° to N. The lowest seam, which furnishes the most of the production of the mine, is very irregular, varying in thickness from one to thirty feet. The anthracite is very friable and flaky.[2]

Spec. gr.	1.86
Parts of lead reduced by one part of coal	27.70
Units of heat	6371.00
Percentage of ash	15.00

At Changkauyü, about eight miles W. by N. from Fangshan, is the Tashhitang mine, which is interesting as showing the manner in which the Chinese work on a large scale. The inclination of the seam varies from 50° to 90°, and the thickness from one to thirty feet, the average being estimated at six feet. The coal is called *haimé*, *i. e.*, black coal, and is a hard, lustreless anthracite, in layers with irregular fracture.

Spec gr.	1.80
Parts of lead reduced by one part of coal	31.50
Units of heat	7245.00
Percentage of ash	5.50

[1] See Appendix No. 2 for better analyses.

[2] See Appendix No. 2.

The workings extend to a horizontal distance of about 6,000 feet, the drainage being effected by a fault, and the ventilation by an opening through old workings to day-light.

The mine is entered by an inclined gallery, descending in the seam, at an angle of about 30°, till near the water level. From the foot of this a horizontal or slightly rising level is driven in the coal to the extreme limit of the intended mine, in this instance over 6,000 feet.

In extracting the coal only those portions of the seam are worked which are sufficiently thick to admit the miner without cutting into the walls.

The "winning" is conducted on the following general plan: where the coal is sufficiently thick, rising galleries are driven at an angle of about 30°, from the tops of which a level extends in both directions as far as the seam retains the proper thickness. From this level other rising galleries and a second level are driven, and so on till the whole enlarged part of the seam is opened, forming pillars twenty-five or thirty feet high, with a length that seems to be very variable. The timbering is now removed from the upper gallery, and the coal broken down from the roof, the miner working from a scaffolding. In this manner working from the farthest and uppermost pillars toward the main level the coal is all taken out, unless the extent of the enlarged part of the seam is too great, in which case pillars are left standing. The coal is all carried on basket-sleds to the main level, and through this to the surface. A great deal of timbering is used, chiefly the wood of fruit trees, etc., and costing at the mine twenty-nine cents per 100 lbs.

One miner produces on the average about 700 lbs. daily, his wages being thirty-nine cents. About four-fifths of the coal is a mixture of small pieces and powder. The owner of the mine considered himself able to produce between thirty and forty tons, of coarse and fine, daily. The price at the mine is $3.60 per ton (2000 lbs.) for the lump coal, and $2.00 for the fine, which is bought to make cakes similar to our patent fuel. The better varieties of the Fangshan coals are taken to a depot at the head of boat navigation on the Liuli Ho,[1] about twelve miles from Fangshan, where the selling price is about $5.50 per ton.

The better varieties of the Chaitang and Muntakau districts are carried on mules and camels to Peking, where the selling price of the former is about two and a half times the price at the mines.

So far as I could ascertain, all the coal worked in the district of Fangshan and in the eastern portion of the Wangping field is anthracite. The only instance of an intrusive rock that I observed in the Fangshan district, was west of the city,

Fig. 6

a. Granite. b. Fine-grained micaceous rock. c. Sandstone altered to quartzite. d. Limestone. e. Black clay-shale with four seams f of anthracite. g. Quartzose conglomerate. h. Creek alluvion.

[1] A tributary of the Peiho.

where a low ridge of granite runs N. S. and is succeeded on its western side by the vertical coal rocks, also trending N. S., while almost everywhere else in the district the strike of these last is E. W.

The preceding section is simply intended to show the relation of the strata to the granite.[1] The limestone *d*, is about 600 feet thick, and seems to be a member of the Coal measures proper. The black shale *e*, with its seams of anthracite *f*, is about 500 feet thick.

From the Plain of Peking to Kalgan.

As we approach the Nankau pass, through which lies the great high-road from Peking to Central and Western Asia, we find the edge of the plain deposit rising with a more rapid slope toward the bordering mountains, while at the same time, the firm, fine loam gives place to rolled fragments and gravel of limestone and granite, from the neighboring hills. The pass is reached by the transverse valley of the Nankau creek.

Leaving the plain, we pass between lofty cliffs of limestone for about six miles, before reaching the axial granite of the ridge. The trend of the strata, which is N. 60° E., with a dip of 40° to S. E. by S. ¼ S. at the edge of the plain, becomes irregular as we approach the granite, the beds being in places almost horizontal, and in others vertical and striking E. W. The latter case occurs at about two and a half miles from the plain, where a side ravine discloses a dyke of a black eruptive rock, inclosed between the strata to which its plane is parallel. This rock has, in a black compact base, thin transparent crystals of amber colored triclinic feldspar. The dyke is only a few feet thick, and is made up of transverse columns. Near the grand marble arch of the Küyungkwan, the limestone is cut through by red porphyry, which is itself traversed by a greenstone dyke. The porphyry contains a little quartz, green mica, and crystals of orthoclase in a compact pink base. The greenstone is apparently a fine-grained diorite.

The granite of the Nankau pass consists chiefly of large crystals of flesh-colored orthoclase, black mica, and comparatively little quartz, with crystals of white triclinic feldspar. Near the middle of the pass there is a different and somewhat remarkable variety, almost free from mica, and consisting of pearly white orthoclase and gray quartz in nearly equal proportions. It is slightly cellular, containing prismatic crystals of white and smoky quartz in the small cavities.

The first of these varieties is traversed near Chatau by dykes of a pink rock, consisting of a fine-grained mixture of orthoclase and quartz with very little greenish mica—one of those rocks that form the link between quartziferous porphyry and true granite. These dykes are in places crossed by others, probably of diorite, consisting of a fine-grained mass of hornblende and feldspar.

The ridge we have just crossed extends to the S. W., forming, in Shansi near the Chihli boundary, a series of high peaks which, on the 26th of April, 1864, were

---

[1] Unfortunately most of the specimens and notes from this interesting locality were lost.

covered with snow, rendering their great domes visible from the valley of the Yang Ho, towering above the mountains that occupy the intervening space of sixty or eighty miles. From the low Nankau pass, we descend to the Kwei Ho, a small tributary of the Yang Ho, which occupies a broad N. E. S. W. valley.

High terraces of a recent lake-deposit occupy the greater part of the valley, concealing the rocks and resting at Chatau on the granite. About a mile west of Chatau rise small hills of a porphyry conglomerate, in beds trending E. N. E. and dipping to N. N. W. about 40°. As we go toward Yülin the fragments and rubble on the surface consist of porphyry, granite, and some limestone.

Descending from the lake terraces and crossing the flats of the Kwei Ho we reach Hweilai (hien), situated on the terrace that fringes the northern border of the valley. Within the walls of this city limestone is seen to crop out in beds trending nearly N. E., and dipping to N. W. Going N. W. from here, over the terrace, the only index to the structure of the neighboring hills is in the angular and rounded fragments on the surface, and these consist of hornblendic gneiss, granite, quartz, porphyries and limestone till Shachung.

Between this city and the town of Sinpaungan the hills consist of the Coal measures, resting on the limestone, which here dips N. W. into the mountains called Papaushan. (See sect. Pl. III.) Between the coal rocks of this mountain and the remarkable limestone hill Kimingshan, there is an anticlinal basin filled with gravels of the lake terrace deposit, and formed by the erosion of an anticlinal fold of the limestone.

In the Kiming mountain the limestone beds are almost vertical, and so highly metamorphosed that in places the rock is almost flint, and their trend has changed to N. S. On the western side of the hill are the vertical strata of the Coal measures with seams of anthracite of poor quality, that have long been worked. The coal rocks of Kiming bend around the northern end of the hill, and extend away to the east, while on the other side of the Yang Ho they seem to extend up the valley of the Sankang Ho.

Crossing this small field to the northwest along the Yang Ho, we reach a deep gorge, through which the river traverses the limestone ridge that forms the northern border of the coal basin. In this gorge the limestone trends N. 70° to 75° E., dipping 25° to S. by E. ¼ E. Near the village of Hiangshui (pu), at the N. W. end of the gorge, the limestone suddenly ceases, and an open country of low hills of a peculiar rock, an amygdaloid, succeeds to the high ridge of limestone. Near the line of contact, the limestone trends as before, E. by N., dipping to S. by E., while the beds of the amygdaloid have the same trend, but a northerly dip. Here we seem to be on the line of an immense fault, for, although the fault itself was not seen, everything seems to point to it. The amygdaloid contains fragments of limestone, and strongly resembles in every respect a similar rock, which we shall see further on, forming a member of the Kiming Coal measures. This slip must have been extensive, as the limestone cliffs seem to be nearly 1000 feet high. The amygdaloid, corresponding apparently to the Schalstein of the Germans, is, perhaps, a tufa of the greenstone-porphyry that occurs in it in fragments.

We soon emerge from these hills upon the plains of Siuenhwa (fu), which occupy another enlargement of the Yang Ho valley, and are also lake terrace deposits. The road lies over this lake bed till about ten miles N. E. of the city of Siuenhwa (fu), where a spur extends westward from the mountains. This spur consists of a double ridge, with an intervening longitudinal depression, the southernmost portion being formed by beds, highly inclined to N. and trending E. W., of quartzite, red argillaceous sandstone, and a compact white rock, apparently an altered argillite. These beds, which seem to be the equivalent of the great limestone formation, will be referred to again in discussing the Hwaingan strata.

The northern part of the double ridge is a remarkable porphyry, which has either traversed or overlies the last mentioned beds. This rock may be called the Kalgan[1] porphyry, as it is extensively developed around that city, although it occurs also in the hills of the Gobi desert. It belongs to the trachytic series.

On the southern flank of this spur the lake deposit rises rapidly toward the hills, and the firm loam, of which it here consists, is cut into by deep gullies. In one of these places a section is exposed of horizontal beds, apparently the tufas of the

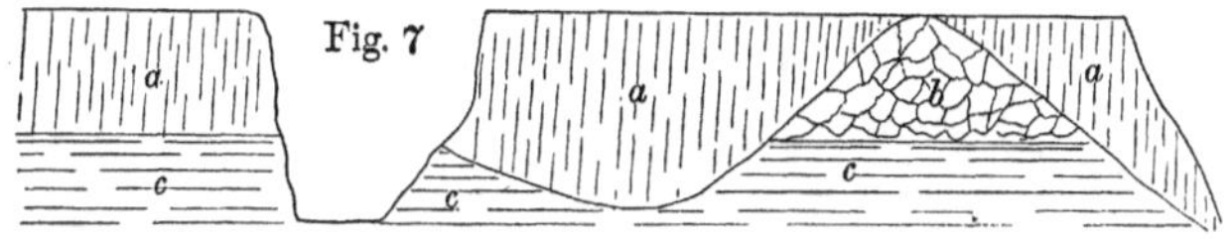

a. Terrace loam. *b.* White tufa. *c.* Red tufaceous sandstone.

Kalgan porphyry. The effects of an erosion previous to the deposition of the lake loam are visible.

We shall find similar tufaceous deposits intimately associated with the Kalgan porphyry near that town.

From the spur we have been examining we follow the road over the lake deposit, to Kalgan, or Changkiakau. High and rugged hills of the trachytic porphyry inclose the valley on the east, while to the north lies a higher range of mountains, which, as it forms a geographical as well as political boundary, and represents approximately the line of the Great Wall, we may call the Barrier range.

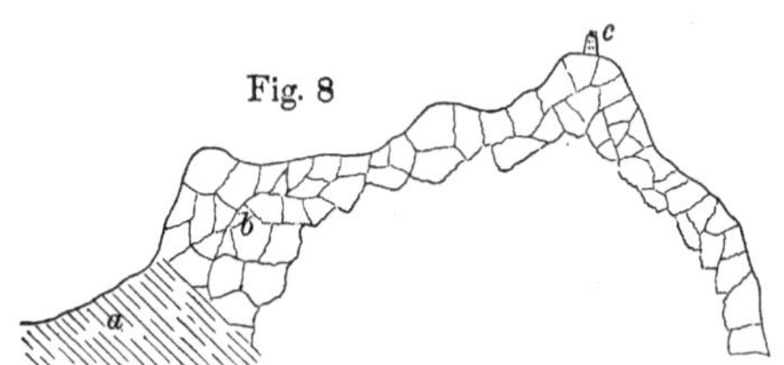

a. White and red tufas. *b.* Kalgan porphyry. *c.* Tower of the Great Wall.

At Kalgan this range is traversed by a gorge, with vertical walls, through which a small stream finds its way to the Yang Ho from the edge of the Mongolian plateau.

[1] The Russian name for Changkiakau, an important market town and gate of the Great Wall.

Here is the most important gate of the Great Wall through which pass all the caravans to Russia, and nearly all those that trade with Western Asia.

The mountains here consist of the tufaceous rocks of the Kalgan porphyry, which are traversed by dykes, and contain beds, of the parent rock. The portions of the range where this formation predominates are easily distinguished from those consisting of the usual granite and metamorphic schists, the latter forming pyramidal hills, while the former have the castellated appearance that is given by cliffs and dykes. The white and red tufas form low hills west of Kalgan, and in the wall of the gorge, in the Barrier range, beds of these rocks trending E. W., and dipping about 45° to N., seem to extend under the porphyry, Fig. 8.

CHAPTER IV.[1]

STRUCTURE OF THE SOUTHERN EDGE OF THE GREAT TABLE LAND, AND OF NORTHERN SHANSI AND CHIHLI.

Two roads, slightly divergent, lead from Kalgan to Urtai on the plateau. About a mile and a half from the town, on the east road, the trachytic porphyry formation appears, under circumstances that would seem to show that much of it is of pluto-neptunian origin.

This formation extends several miles further north and northeast till it is limited by the metamorphic schists of the range. On the west road the same formation exists till near Tutinza, on the northern side of the range, and furnishes slabs of tufa and blocks of porphyry for building purposes.

The country crossed by the road between the Barrier range and the edge of the plateau is a depression, here about nine miles broad. On either side of the road are flat-topped hills 80 to 100 feet high, of gravel made up in great part of rolled fragments of quartziferous porphyry. This gravel, which I take to be of the same age as the lake loam and terrace deposits, also forms the low hills traversed by the eastern road, where it covers a brown-coal basin probably of tertiary origin, of which, unfortunately, I was able to see only specimens of the coal.

About half way between Tutinza and Hanoor the road begins to rise to the plateau, and leaving China proper, with the edge of the table-land, we reach the steppes of Tartary.

The height of the edge is here 5,400 feet above the sea, according to the measurement of Fuss and v. Bunge, and probably not less than from 3,000 to 3,500 feet above Changkiakau, and the edge itself forms a precipitous wall to the south, while the plateau slopes off gently to the north.

From a tower of the Great Wall, which crowns a hill near Hanoor, we have, spread out before us, a grand panorama of the surrounding country. The natural wall formed by the abrupt termination of the table-land stretches away from the tower far off to the west and northeast, bounding the valley south of it as a precipitous coast bounds the sea. Between us and the Barrier range, the depression, occupied by low hills of the eroded gravels, lies like a neutral belt between two regions of the earth in almost every respect widely different each from the other. To the south only barren and rugged mountains meet the eye, and beyond these to the Southern Ocean, the mountainous character is redeemed only by the fertile valleys of a few

[1] For this Chapter see Map, Pl. No. 2, and Sections, Pl. No. 3.

large rivers. To the north lie the endless plains of Tartary rarely crossed by other than low ridges.

At the point where the road begins to rise to the table-land, we enter upon the volcanic formation of Southern Mongolia. From the base of the plateau-wall to the summit, we may look in vain for other than the rocks of this formation, and as we travel westward we shall see little else while on the plateau.

Our road now follows a general westerly course, keeping near the edge of the table-land. The surface of the plateau along this route is everywhere cut into by valleys varying in depth from one to several hundred feet. The tops of the hills thus formed are flat, and in the same plane—that of the original plateau surface—excepting where the erosion has isolated small hills, in which case they present knobs lower than the general plane. The sides of these hills form in places cliffs, but more generally they slope off to the valley bottoms. The width of the valleys varies from a few hundred feet to three or four miles, the smaller ones sometimes narrowing to a gorge, and again reopening to their usual size. They frequently form fertile meadows with brooks winding through them, and are then the camping grounds of the Mongols, and the pastures of their large herds of sheep, horses, cows, and camels. The pasture is not confined to the bottoms, the whole country, hill and valley, being clothed with excellent grass.

Soon after leaving Hanoor we reach a small lake, or rather pond, without outlet, inclosed in the depression between several knobs. It is difficult to understand how these small depressions are formed, unless we suppose them to represent former inequalities in the bottoms of valleys once occupied by running streams. Such small lakes are characteristic of Mongolia, and we shall have occasion to notice several.

Continuing westward, the road passes the lama-monastery of Boroseiji, and ascends the grassy valley of a small tributary of the Narin Gol.[1] This stream rises at the very edge of the plateau, flows N. E. by Urtai, and turning to the south descends from the plateau at Teutai, and passing through the gorge at Changkiakau, joins the Yang Ho.

Leaving the system of this stream, we pass over a ridge, part of the original plateau, near which is a hill rising several hundred feet above us, consisting, to judge from fragments on the surface near by, of chloritic gneiss. This is an isolated peak, rising through the volcanic formation which has buried the rest of the ridge.

Descending to the west we enter another fine valley, apparently that of a tributary of Angouli Noor.[2] Through this valley flows a creek which, near the Mongol village of Hanoortai, widens to a small lake, the abode in summer of thousands of wild ducks. From this valley the road passes over a low ridge and descends by a narrow, rocky defile to the plain of Taulichuen, in which is the source of one of the tributaries of the Yang Ho. We have here left the plateau, and are among the cultivated fields of the Chinese,[3] but we are still on the volcanic formation.

---

[1] Gol, Mong. for river. Wherever this word occurs in this itinerary it refers only to small brooks.

[2] Noor, Mong. for lake.

[3] The Chinese are forbidden by law the cultivation of land on the plateau.

Leaving this plain, we again rise to the table-land, and following, for six or seven miles, its abrupt edge we come again to a sudden descent by which we leave it and enter upon a rolling country. The plateau wall makes here a great bend, trending away to the northwest.

The country over which our road now lies is a rolling plateau formed by a broad swell, or ridge, of the granitic and schistose rocks, from which the volcanic plateau covering has been eroded. On it are the sources of another tributary of the Yang Ho.

The rocks are granite, syenite, and crystalline metamorphic schists.

This bay-shaped indentation of the southern edge of the plateau is about 15 or 20 miles broad; it is drained in part by a valley descending toward the southwest, and is surrounded on the east, west, and north by the wall of the higher plateau. The northern portion of this bay forms a depression that is only partially drained, and which at times is evidently a marshy region, while it contains at all seasons three small lakes—Gurban Noor. In April the country about these lakes was covered with scattered tufts of grass, between which the dry clayey surface was white with an efflorescence of soda, and the borders of the lakes also were incrusted with a dazzling layer of the same salt.

About two miles west of the Mongolian camp of Gurban Noor, the higher table-land again begins, but with a somewhat different character. Rising to the top of a granite ridge, we descend a little on the west into a plateau-valley. On either side and before us are everywhere the same flat-topped hills we have seen forming the table-land, but they are only the remnants of a volcanic covering insignificant in thickness compared with that we have seen farther east. The valleys have everywhere cut through this covering and into the granito-schistose foundation.

Our road now lies through a succession of circular and oblong meadow-valleys, connected by narrow outlets, thus forming one valley-course, and containing a small brook, the Hoyurtoloho Gol, which flows S. E. The meadow enlargements are evidently the beds of small lakes filled with the detritus of the surrounding volcanic and granitic rocks.

Following this valley in a general S. W. direction from the Mongol camp, Hoyurtoloho Gol, we descend through a narrow defile in chloritic granite, into another bay cut out of the plateau, and open to the S. E., where the drainage finds an exit through the valley of the Si Ho, another tributary of the Yang Ho.

Soon after leaving the gorge, by which we have descended, the road crosses a lava stream one or two thousand feet broad, and from sixty to eighty feet thick, which crosses the valley, and is cut through by the rivulet. In this section it shows columnar structure, and is in places porous and amygdaloidal. A mountain forming apparently a detached portion of the neighboring plateau, and having the appearance of a half-destroyed crater, seems to be the origin of the stream. The eruption causing this occurrence must have been subsequent to the erosion of this part of the plateau, and was probably subaerial. The locality is interesting as being the only one in which I noticed traces of true volcanic action more recent than that to which the volcanic formation of Southern Mongolia owes its origin.

Crossing the valley of the Si Ho, which leaves this bay-shaped depression at the

S. E., we enter another valley opening in the S. W. Frequent fragments of a calcareous deposit strewed over the surface indicate the action of mineral springs.

Gradually ascending this valley, which, as well as that of the Si Ho, is occupied by a deposit of loam, probably contemporaneous with the terrace loam of the Yang Ho, we reach a point where this loam deposit, by forming a bar across the valley, causes a low watershed, on one side of which any drainage there may be flows north to the Si Ho, and on the other south to the undrained lake Chaganoussu.

We shall see that this remarkable occurrence of alluvial watersheds stretching across valleys is intimately connected with the formation of the undrained lakes of this portion of Mongolia, having its origin in a former system of great inland lakes, and its continuance in the dryness of the climate.

The grassy valley of Chaganoussu has two other openings through the plateau, one on the east connecting it with the Si Ho valley, and another on the west leading to the Kir Noor. Both of these are crossed by bars covered by the terrace loam, if not entirely formed by it. Our road, after skirting the shallow pond of Chaganoussu enters the valley leading to the southwest, and passing the dried up bed of the Hoyur Noor descends through a narrow defile till it emerges into the great depression of the Kir Noor.

From the Si Ho to this point the rocks, both of the adjoining plateau and of the exposed parts of the valley bottom, belong throughout to the volcanic formation.

From the edge of the plateau, near where the road enters the Kir Noor valley, a view of the whole of this ancient lake-bed is spread out beneath us. It is a large plain about 15 miles broad, its longer axis trending about N. N. W. On both sides the lofty and bold plateau edge is seen stretching away to N. N. W. and S. S. E., as far as the eye can reach, without meeting to inclose the valley.

Away to the southwest of us a distant portion of the plain covered with a dazzling white efflorescence marks the position of the Kir Noor of a few years since. From this, the most depressed part of the plain, the surface rises toward every point of the compass. Far away to the north a bar of the lake deposit seems to stretch from wall to wall of the valley, while in the south this is certainly the case. Over this southern alluvial bar the peaks of the Barrier range are seen in the distance.

To the N. N. W. a distant peak, capped with snow (April 18th), is visible rising above the level line of the table-land.

The edge of the plateau on both sides of the valley, wherever I visited it, consists of the volcanic formation, from the summit to under the lake deposits, but the presence on the surface of the latter of granite detritus indicates the presence of the older rocks at no great distance.

East of the Mongol village of Hoyurbaishin, a gully exposes a section of the plain deposit near where this abuts against the edge of the plateau. The deposit is stratified, and its beds have the same dip as the surface of the plain. It consists of coarse sandstones and fine conglomerates, formed from the detritus of the neighboring volcanic rocks and cemented by a calcareous mineral, the product, perhaps, of springs, which enveloping each grain or pebble with concentric layers produces a hard rock. The only trees seen in the valley of the Kirnoor were two old ones growing in this gully, nor did we meet with any others either on the plateau or in its valleys.

The lake is said to be drying up, and the Mongols say that its waters have flowed into the Té Hai farther west, an apparently unfounded belief, as there is no surface communication between the two lakes, and the natives on the shores of the Té Hai were not aware of any increase in its volume. Still it is evident that the waters of the Kir Noor are rapidly disappearing, and the cause, whether this be only temporary or a constantly operating change in the climate, has been acting for at least several years. Among the lakes we have already noticed, the Chaganoussu is also disappearing, and the adjoining Hoyur Noor has for several years been represented only by its dry bed.

The greater part of the plain of the Kir Noor valley is clothed with grass, and supports large herds of sheep, but as we approach the recent lake-bed the surface is eroded by dry, shallow water-courses, and is covered with tufts only of grass, between which the ground is bare and cracked. This was apparently a marsh surrounding the lake of which, a little further west, the dry bed is visible covered with the white soda efflorescence, and stretching several miles west, north, and south.[1]

The walls of this great valley, formed by the abrupt edge of the plateau, are marked by a series of lines at different heights, and extending apparently horizontally, and on the same level, along the faces of both sides of the valley. They are reproduced on an island-like hill that rises from the plain, and are visible at a distance of from ten to twelve miles to the naked eye. They are defined, where the slope is gentle, by a continuous mass of large and small fragments of rock, and on the steep declivities by slight variation in the angle of slope.

I was able to examine these lines in only one locality, and there they appeared to be independent of the structure of the plateau, and I can account for them only on the supposition that they mark former water levels.

Following the road from Hoyurbaishin to the Té Hai we cross, at about the middle of the valley, a small stream of fresh water flowing from the north, and which is seen to empty into the remnant of the lake a mile or two south of the road. Still farther west the road lies through a marshy tract. Two or three miles west of this we reach a terrace of the lake-deposit, which descending rapidly from the western side of the valley, faces the plain with a bluff. As the road ascends a ravine in this terrace, the increasing proportion of fragments of granite and gneiss shows that we are in the neighborhood of a rise in the granite foundation, while a few miles to the north a ridge rising several hundred feet above the level of the plateau, seems to be the source of the fragments in question.

As we leave the terrace and the valley of Kir Noor, we pass a deep and gloomy gorge cut through the plateau to its very foundation. Where seen it is barely separated by a low ridge from a valley that leads into the Kir Noor. This chasm seems to lead to the Karaoussu, a tributary of the Tourgen Gol, which is an affluent of the Yellow river. The valley by which we leave the plain leads us in a S. S. W.

---

[1] For the results of an examination of the dried mud of the recent lake-bed, see Nos. 1 and 12 in Mr. A. M. Edwards' Letter, Appendix No. 3.

direction gradually ascending, the flat-topped hills of the table-land shutting us in on both sides, till we reach a watershed from which we look down on a large, deep, circular valley, covered with grazing herds, and ornamented with the gilded spires of a lama-temple. This valley is shut in on the north and west by the volcanic formation of the plateau, but its southern wall is of granite and garnetic gneiss, capped here and there by thin remnants of the plateau mantle. Still farther south, after passing the village of Yingmachuen the plateau formation predominates, and the long descent into the valley of the Té Hai[1] is entirely over its rocks.

The great depression of the Té Hai is about twelve miles broad, and so far as the plateau is concerned, appears to be open to the S. W. in the direction of its longer axis. The northwestern side is formed by a serrated range of mountains, which rises about 2,000 feet above the lake, between this and the plateau. The eastern wall is of gneiss capped with the volcanic plateau formation, and the same would seem to be the case with the southern wall, while, as we have seen, the northeastern side is volcanic in its entire height. Thus the thickness of the volcanic mantle varies, within a few miles, several hundred feet.

The northeastern end of the valley contains an extensive deposit of the terrace loam. This faces the lake with a bluff that stretches N. W. S. E. across the valley.

From this line the terrace rises toward the N. E. at first gradually, and then rapidly, until in the long northeastern arm of the valley and in the side valleys, its surface is several hundred feet above the lake.

Below this terrace a plain rises gently from the lake toward the mountains.

The terrace deposit is a firm, stratified loam, containing, near the hills, numerous fragments of the neighboring rocks and layers of gravel. It is cut into by deep ravines, in the sides of one of which, about five miles east of the lake, I found several species of fresh-water univalves.

The lake is apparently about eight miles long by four or five broad. Its water is salt, though far less so than seawater, and is not bitter. The flat surrounding it is covered with a thin coating of soda efflorescence.[2]

While the valley of the Kir Noor is occupied exclusively by the Mongols and their herds, that of the Té Hai is cultivated by Chinese, only one or two Mongol camps being seen. Ancient watch towers, that dominate these plains, and from which signals could be made to the long line of similar posts on the Great Wall, are silent monuments of a time when the shores of these lakes were the home of an aggressive race, ever threatening a descent into the fertile regions of China. Rising with the terrace, the road leads us to the hills that form the southeastern wall of the valley, and we pass through these by a deep and rocky ravine, in which the pass is situated. These hills are, as I have already said, of gneiss, characterized by an abundance of garnets, and capped with the volcanic mantle. The stratification trends, in the main, N. E. and dips 75° to N. W. Garnetiferous granulite, from these

---

[1] Daikha Noor of the Mongols.

[2] For negative results of a microscopical examination of the deposits, both of the terrace and the flats, see Nos. 2 and 3, in Mr. A. M. Edwards' Letter, Appendix No. 3.

hills, occurs in the terrace deposits on their N. W. flank. From this hill we descend into a small valley which empties into that of the Té Hai. In this valley the terrace loam is present to the height of probably not less than 250 feet above the lake.

From here the road descends to the deep channel cut through the plateau, which connects the great valley of the Té Hai with that of the Sankang Ho. This channel is cut to the bottom of the volcanic mantle, here apparently over 1,000 feet thick, and into the metamorphic rocks on which it lies.

In this channel we meet with another of those remarkable watersheds of terrace deposit which stretching from wall to wall, slopes on the west toward the Té Hai, and on the east toward the valley of the Sankang Ho. The material forming this bar is almost loose sand mixed with fragments from the volcanic and metamorphic rocks, and is but little, if at all, eroded on the western flank, while there are gullies on the eastern in which highly inclined beds of granulite, containing garnets, are exposed.

At Maanmiau the valley opens to form the broad, swampy plain of Fungching, rising from which are frequent low hillocks of gneiss in strata trending between E. and N. E. Here the high plateau leaves the road; the part that has formed the southern side of the valley since leaving the Té Hai, now trends away to the S. S. W. till the steep face and level outline of its edge are lost in the far distance. On the other side, the part which has formed the northern wall of the valley, continues a few miles farther, and then, before reaching Fungching, bears away to E. N. E.

Although we have here left the higher plateau, we have not yet reached the southern limit of the volcanic formation. At a level of perhaps 1,000 feet below the surface of the higher plateau begins the lower plateau, the flat surface of which is 200 or 300 feet above the valley, and extends southward from the very edge of the higher. It consists of the same volcanic formation as the higher table-land of which it was, I think, without doubt, once the continuation, the continuity having been broken by an immense fault—a supposition to which I shall recur further on.

The marshy plain of Fungching is fringed in places with low, flat hills, which owe their form to the terrace deposit of loam, but under this, consist of a bright red, sometimes loose material, apparently a wacke or a product of the decomposition of the volcanic rocks. In this are fragments of a red calcareous mineral, a product of the action of waters on the adjoining rock before or during its alteration. We shall see a similar mineral filling crevices in the volcanic plateau formation. It is perhaps the result of the metamorphic action of mineral springs rising along the great fault-line.

A few miles beyond Fungching our road rises to the surface of the lower plateau, and we obtain an open view from a ruined part of the Great Wall. To the north we can see the precipitous edge of the higher table-land stretching far away to the northeast, the break in it formed by the valley of the Kir Noor, and its continuation beyond this toward the Si Ho.[1] To the south and east we see the barren crest and peaks of the Barrier range. Between the higher table-land and this sierra is the lower plateau on the southernmost spur of which we are standing. The valley we

---

[1] In Mongol, Djookha Gol.

have followed from the Té Hai passes beneath us, and continues south to Tatung (fu) and the Sankang Ho; it is well watered and fertile.

Crossing this southern promontory of the lower plateau the road descends into the valley of Kwantung (pu), a depression occupied by another tributary of the Sankang Ho, and lying between the lower plateau and the Barrier range. This range and a spur from it, form the southern and eastern limits of the valley, and the lower plateau forms the northern side, while to the west it is open.

A quarry about half way up the edge of the plateau presents a good though limited section in the volcanic formation. In this quarry two beds are visible—a lower one of crystalline lava, which, toward the top, becomes porous and passes into a true scoria, and an upper bed of more compact lava. Crevices extending through both these beds are filled with a calcareous segregation.

The terrace deposit sweeps from the valley of Fungching around the southern spur of the lower plateau, into the valley of Kwantung, from the centre of which it rises rapidly up to the sides of the mountains, filling their ravines, to a height of several hundred feet above the middle of the valley.

From the mountains forming the northeastern side a low spur juts out, narrowing the valley, and in the space between the point of this spur and the southern wall of the valley there is another of those remarkable watersheds to which I have several times alluded. The terrace deposit rises from the west to form this bar (though without reaching a height at all comparable to that to which it rises on the mountain sides) and falls off again toward the southeast.

Crossing this bar, and descending toward the southeast, we traverse the Barrier range by a deep and narrow gorge about eight miles long, through which flows a small stream which, taking its rise in the northeastern part of the valley of Kwan tung, empties into the Yang Ho.

In this gorge the range is seen to consist of crystalline metamorphic schists, chiefly gneiss, hornblende gneiss, hornblende schist, and hypersthenite, in strata varying in trend between N. N. W. and N. N. E., the dip at the two ends of the defile being toward the centre.

The terrace deposit occurs in this gorge and its side ravines, high above the stream, and on emerging into the great valley of Yangkau it is seen rising from the plain with an unbroken surface high up the sides of the Sierra north of the Yangkau valley, while south of the mouth of the defile it exist only as terraces several hundred feet above the plain. The terrace deposit extends from here down the valley of the Yang Ho to form the plains and terraces of the enlargements of the valley at Siuenhwa (fu) and Shachung. But it is not confined to the present river systems, for east of Tienching (hien) it caps the lower part of the ridge between the valleys of Yangkau (hien) and Hwaingan (hien) forming a plateau of loam several hundred feet above the valleys.

Following the road from Yangkau to Tienching, we have on the north the Barrier range, a rugged sierra of which the barren peaks must be from 2,000 to 3,000 feet high, above the valley. Along the line where the terrace deposit terminates on the steep flank of the sierra, extends the now ruined Great Wall of China, with its towers and parapets, till at a point opposite Tienching it crosses the mountains to extend northward to the high plateau. The southern side of the valley is formed by a lower ridge, beyond which higher mountains are seen, and over these the distant snow-capped[1] peaks, or rather domes, of the range south of the Sankang Ho.

Leaving the valley of the Yang Ho near Tienching, we cross over the terrace-capped ridge before mentioned, into the valley of Hwaingan (hien). To the north of the road in crossing, and north of the whole valley of Hwaingan, the hills are seen to consist of alternating strata of a bright red rock and of a harder rock, in anticlinal and synclinal folds. The fragments brought by streams from the hill forming the western part of the southern side of the valley, are gneiss and hornblende schist.

Following the Hwaingan creek to the northeast, the road approaches, near where it emerges into the valley of the Yang Ho, a fine section in the strata of the northern hills. Resting on gneiss are strata of highly metamorphosed rocks, the continuation of those we saw in the hills between Siuenhwa (fu) and Kalgan, and which for the present may be called the Hwaingan beds. The valley of Hwaingan trends N. E. by E., and this seems to be about the strike of the strata. In the exit into the valley of the Yang Ho, the Hwaingan creek flows through a gorge formed by the erosion, parallel to its axis, of an anticlinal ridge of the Hwaingan beds.

From this point our road crosses the valley of the Yang Ho, and brings us again to Kalgan.

KALGAN TO SIWAN AND SINPAUNGAN.

Leaving Kalgan the road runs in a northeasterly direction through a deep gorge, with vertical walls, in the Kalgan trachytic porphyry, and its pluto-neptunian deposits, as far as Ulanhada. At this village it leaves the valley of the main stream, and turning into a tributary valley, winds with this through the mountains, following an easterly course to the Roman mission of Siwan. For eight or ten miles we see only the rocks of the Kalgan porphyry, but before reaching the village of Siyin'sz, these are followed by the crystalline metamorphic schists, which in turn are succeeded, before we reach Siwan, by syenitic granite. This last is eruptive, dykes of it traversing the metamorphic strata, and the main body often containing fragments of the schists. This rock forms the mountains around and beyond Siwan.

From Kalgan to this point, and beyond, the terrace deposit occupies the sides of the mountains, and at Siwan its terraces form the sides of the valley to the height of from 200 to 300 feet above the creek, and its vertical cliffs show it to be a fine, compact loam. In it the Chinese excavate their dwellings in suites of apartments having doors, windows, and partition walls, all cut in the loam. The walls are simply plastered over to prevent the dust from falling, and in this condition they last as long, if not longer, than the ordinary houses built of sunburnt clay.[2] In the

---

[1] 26th April, 1864.

[2] These excavations are common wherever the terrace deposit occurs in Northern China.

course of these excavations, fossil remains of quadrupeds are obtained in considerable numbers, especially horns of deer.[1]

Leaving Siwan the road lies first southeast, then south, crossing two ridges of chloritic gneiss and chloritic schist, and descending into the large oval valley of Chauchuen. This valley is occupied by the terrace deposit. Our road ascends the ridge forming the southern side of the valley. On the northern flank are the crystalline metamorphic schists covered by limestone, and over this beds of porphyry breccia with dykes of eurite. The terrace deposit rises almost to the summit of this ridge on both sides. Descending through the deep gullies in the terrace loam, the road enters the valley of a creek that empties into the Yang Ho, just north of the Kiming mountain. From this valley we cross the ridge, by a low pass east of the Kiming mountain, into the valley of the Yang Ho, and descend to Sinpaungan. The low pass is covered by the terrace deposit, and beneath this on the northern flank are the coal rocks of the Kiming field, among which I saw a greenstone porphyry conglomerate similar to that at Hiangshui (pu), and probably its equivalent.

The terrace deposit in the pass consists of loam with gravel and fragments of the neighboring rocks, and occupies a higher level than the terraces of the valley to the south.

I will now attempt a general description of the principal rocks met with on the above journey. I am well aware that the following description can have but a very limited value, owing to the absence both of chemical determinations and of closer observations of the modes of occurrence.

Granitic and Crystalline Metamorphic Series.

Distribution.—These two classes of rocks form either collectively or individually the main body of every ridge we have traversed. Of them consist the ridges that rise through and above the volcanic mantle of the plateau, and they form the foundation on which this rests wherever the foundation was seen. Indeed, they are the skeleton of this region, supporting the limestone floor of the coal rocks.

Granite predominates in the first range where we crossed it in the Nankau pass; in the other localities, if it exist, it is covered by the crystalline schists.

Unstratified Granitic Rocks.—The main body of the ridge between Nankau and Chatau consists of a granite containing two varieties of feldspar, about equally distributed in crystals varying from an eighth of an inch to three-quarters in length. These are pink orthoclase and a white triclinic feldspar. The mica is a dark green almost black, probably magnesian variety, and quartz is present in comparatively small quantity. It is thus a granitite.

Near the middle of the pass is another variety, of even grain, consisting of only white orthoclase and gray quartz, the latter often in sharply-defined, small prismatic crystals imbedded in the mass. It is somewhat remarkable from small cells in which

---

[1] As all the fossils of any value had been sent to Paris previous to my visit, I was unable to obtain any that were worth examining. It is to be desired that those now in Paris will be determined and described in order to fix the age of the terrace formation.

ends and corners of small crystals of the constituent feldspar and quartz are sharply developed.

The hills immediately surrounding Siwan, in the Great Wall range, east of Kalgan, consist of a reddish-gray syenite composed mainly of orthoclase, some gray triclinic feldspar, crystals of hornblende, and a little quartz. Large crystals of orthoclase render it porphyroid. Near the contact of this rock with the crystalline schists west of Siwan, dykes of it are seen in the latter, while fragments of the schists inclosed in the main body of the syenite are additional proof that it is eruptive, and younger than the metamorphic schist formation. Fragments of this syenite are inclosed in the pluto-neptunian rocks of the Kalgan porphyry.

A syenite of medium grain, composed of slightly pink orthoclase and hornblende, occurs over a large part of the rolling land east of Murkwoching.

Fragments of a fine red granitite occur in the bed of the Yang Ho near Kiming, and blocks of a red rock composed of fresh, bright-red orthoclase and grains of a soft, talcose or steatitic mineral, thus approaching a protogine, are common in the Hwaingan creek. At this latter locality there are many fragments of a rock, consisting entirely of a coarsely crystalline, triclinic, feldspar, apparently labradorite, of a grayish tinge tending to blue and weathering white. It contains scattered crystals of a mineral resembling sahlite.

Crystalline Metamorphic Rocks.—The tilted and folded strata of these rocks form for the most part all the ridges we have passed over after leaving Chatau. In the hills northeast of Shachung are beds belonging to the chloritic series—white triclinic feldspar, quartz, chlorite, and magnetic iron—a variety of chloritic gneiss.

In the hills traversed by the road from Kalgan to Siwan, and south to Chauchuen, the predominating rocks are still those of the chloritic series. In the hills south of Siwan I observed chloritic gneiss—orthoclase, chlorite, and quartz—and schist of nearly pure chlorite. In the mountains between Kalgan and Siwan, another well-defined variety of chloritic gneiss occurs, in which the feldspar is, in great part, triclinic. Schists of the hornblendic series also play an important part in this region. They are composed of a greenish-white triclinic feldspar and hornblende, sometimes one of these minerals predominating, sometimes the other. The trend of the uplifts in this region, though irregular, seems to lie between N. and W.

Under the Hwaingan beds near Kiu Hwaingan, the metamorphic schists here represented by gneiss, lie with a remarkable approximation to conformability with these younger strata. This gneiss consists of orthoclase and quartz, and is very poor in mica, excepting on the surface of the slabs into which it breaks.

The Barrier range, where we cross it west of Yangkau, is formed mainly of schists of the hornblendic series. Among these are extensive strata of a rock composed of black hornblende, with strongly defined prismatic cleavage, abundant garnets, and a little white feldspar. Another rock occurs among these strata composed of a greenish-white triclinic feldspar associated with a little black mica, quartz, and hornblende.

The substructure of the plateau, southeast of the Té Hai, is of granulite and gneiss. The former rock is in places fine grained and schistose with minute garnets, but occurs more generally with a coarser structure, in which it is seen to consist of white orthoclase and thin lenticular plates or bands of gray quartz, with abundant irregular grains of garnet of the size of a pea.

The gneiss of this locality runs through several varieties, all alike rich in garnets. Gneiss with garnets is also exposed under the volcanic beds at Yingmachuen, northeast of the Té Hai.

Thus where we cross the Barrier range west of Yangkau, we find the predominating schists to be of the hornblendic series. In the echelon to the east, between the Yang Ho and Hwaingan creek, the schists, that underlie the Hwaingan beds, are mainly of the micaceous series, gneiss being most common. The schists that are exposed west of the Barrier range, between this and the Té Hai, and at Yingmachuen, belong, as we have seen, also mostly to the micaceous series, gneiss predominating and alternating with its congener—granulite. The general trend of the uplift of these latter schists, in the region between Kiu Hwaingan and the Té Hai, is northeasterly and parallel to the course of the Barrier range, while the mean strike of the schists of the hornblendic series, in the main body of the range, seems to be north-northwesterly.

If we glance at the metamorphic region east of Kalgan, we find that its schists belong to the hornblendic and chloritic series, and here also the mean strike seems to lie between north and west.

Have we here to do with the metamorphosed strata of two distinct periods? It would be hasty to assume that such is the case in the absence of more data, but it does not seem improbable that the schists of the hornblendic and chloritic series represent deposits of an earlier age followed by N. W. S. E. foldings of the strata, while the gneiss and granulite series belong to a later epoch which was followed by the N. E. S. W. disturbance.

Hwaingan Beds.—These strata, which have already been referred to as resting almost conformably on gneiss, cover the hills on both sides of the Hwaingan creek, and occur with an easterly trend and northerly dip at the edge of the hills, N. W. of Siuenhwa (fu). They are made up of layers of compact and hard, gray silicious limestone, with quartzose sandstones, red and gray argillites, and quartzite. The predominating rock would seem to be the limestone. The aggregate thickness is several hundred feet. The lowest layers are, first, and resting on the gneiss, a fine grained sandstone, green from thin layers of a green mineral; over this, sandstone altered to quartzite; on this a red argillaceous shale; finally, silicious limestone containing numerous thin layers of chert. The alternating beds at the bottom of the series vary in thickness from six inches to many feet, and in the cliffs seen from the road, I noticed that they frequently thin out and dovetail into each other, an occurrence that seems to indicate frequently changing conditions of level and material.

The Hwaingan beds appear to be the equivalent of the great limestone floor of the coal-bearing rocks, and their character and thinness would seem to indicate that they were formed on the borders of the sea in which that great formation originated. The limestone of the Kiming basin is highly silicified, and its thickness seems to be much less than that of the same formation where it rises from beneath the great plain.

Greenstone-Porphyry Conglomerate.—The beds of this rock were noticed near

Hiangshui (pu), and also in the coal field of Kiming, where they occur apparently as members of the coal-bearing series, and at a higher level than the lower coal seams.

The fragments of porphyry that form the characteristic feature of this deposit, have a base that varies in texture, from compact to finely crystalline, in color from dark reddish-brown to black, and that effervesces slightly in dilute muriatic acid. It contains numerous thin, oblong crystals, of a white triclinic feldspar, from one-eighth to three-quarters of an inch long. Through the base are scattered grains of a white mineral, apparently a zeolite, and scales of what seems to be ichthyophthalmite.

In places, these fragments make up the greater part of the deposit, and it is then difficult to distinguish the inclosed from the inclosing rock. In other places the blocks are scattered through a finely crystalline, dark reddish-brown rock, that is irregularly impregnated with a carbonate, and about as hard as compact limestone. It contains also pieces of an amygdaloidal rock, the cells of which are filled with calcite and a white zeolite; blocks of limestone are also found in it.

The general appearance and manner of occurrence of this deposit suggests the idea that it is of pluto-neptunian origin, and perhaps contemporaneous with the eruption of the greenstone-porphyry. I will add that I did not meet with dykes of this porphyry.

Kalgan Trachytic Porphyry.—This rock, and its pluto-neptunian deposits form the hills around Kalgan, and those that, extending S. E. from that city, send out a spur to the west crossing the road from Siuenhwa.

The porphyry in question is very variable in color, the most common variety being brown, but all shades occur from pitch-black to white, red, and green. The texture of the rock is compact, often almost vitreous, but in structure it ranges from the solid rock of the Kalgan mountain to the cellular and often almost pumiceous variety of the spur between Kalgan and Siuenhwa.

Crystals of white, transparent orthoclase, or glassy feldspar, are always present, and are generally so limpid as to take the color of the variety in which they are imbedded. Small grains of pellucid quartz occur more rarely, but seem in places to belong to the primary ingredients, though they are generally secondary. Mica and hornblende are always absent.

The cells are sometimes long-cylindrical, but more generally flattened, though lying in the same direction. They are filled with different varieties of quartz, as cornelian, chalcedony, and a black silex. More rarely they are filled with calcite.

The base of this rock fuses easily before the blowpipe to a white vesicular glass on the edges.

In intimate connection with this porphyry are strata of a deposit which, from their character and manner of occurrence, appear to be of pluto-neptunian origin, and were probably formed contemporaneously with the eruption of the porphyry. These consist chiefly of a tufa, varying in color from white and gray to purple, and in hardness between that of chalk and limestone. Its texture is rough and earthen in appearance. Through the mass are scattered crystals of glassy feldspar, grains of limpid quartz, and hexagonal scales of dark-brown mica.

Beds of another rock occur, of brick-red and brown colors, and having an earthy base, with small, brilliant crystals of glassy feldspar and grains of pellucid quartz, and inclosing small fragments of other rocks.

This deposit is visible on the southern flank of the spur between Kalgan and Siuenhwa, underlying the terrace loam in horizontal beds (Fig. 7).

At the base of the high hill north of Kalgan the tufa beds are seen to dip under the porphyry at an angle of about 45° (Fig. 8), and trending west they form a series of detached hills. On the roads leading to Tutinza, Teutai, and Siwan, they are traversed by a perfect network of dykes of the porphyry, which rock also caps the summits of the hills, its vertical cliffs and outstanding dykes giving them a bold and castellated appearance.

Although no analyses of these rocks have been made, there is, I think, little doubt that we have here to do with a trachytic porphyry and its tufas.

Volcanic Formation of the Plateau.—The southern elevated edge of the Great Plateau is formed, between the 112th and 115th meridians, of an immense lava bed. How much further it extends beyond the limits given above, or how large its breadth may be toward the north, is unknown; I have only tried to indicate on the map the region which I observed it to occupy. Its breadth is, in places, not less than forty miles, and this may be only a fraction of the real width.

The thickness of the formation is, necessarily, very variable as it fills the inequalities of what was once a mountainous country. At Hanoor it seems to be not less than fifteen hundred feet thick, and the same may be said of it in other localities visited, while we have seen it in places represented by only a thin sheet, covering the metamorphic schists, where these rise to near the surface.

The rocks of this formation may be classed under two types—the one basaltic, the other trachytic.

The basaltic rocks were observed more particularly near Hanoor and to the N. E. of that place. Both compact and finely crystalline varieties occur. They are generally, especially the latter variety, poor in olivine and contain here and there crystals of basaltic hornblende.

At many places in the neighborhood of Hanoor, fragments of a cellular variety occur on the sides of the valleys, in a manner that would seem to indicate, that there is a horizontal bed of it, marking the plane of contact between two flows of lava.

The rocks of the other type are throughout crystalline, though often the texture is very fine, and are generally porous. In color they vary from black to dark gray, while some varieties, especially when weathered, are light gray. In some instances hornblende, or augite, enter abundantly into the composition of the rock, but more generally it seems to consist almost exclusively of white or yellow, triclinic feldspar with greasy lustre, partly in tabular crystals, partly massive. Scattered through this mass are minute specks or grains of a dark to light green mineral, with glassy lustre and conchoidal fracture, harder than the knife when fresh, soft and resinous in lustre when altered. The feldspar is probably oligoklas. A characteristic feature of the different varieties of this rock is the extreme rarity or total absence of magnetic iron.

This lava seems to belong to the trachydoleritic series. Of its varieties consist nearly the whole of that portion of the volcanic formation that was traversed by my route. That it obtained its great development on the surface by successive flows, is evident from the stratiform structure of this part of the plateau.

The only locality in which I observed an exposed section of comparatively fresh rock, was in a quarry at Kwantung (pu), on the lower plateau. Here a bed of lava, crystalline at the bottom of the section, becomes porous toward the top, and, finally, highly vesicular and highly scoriaceous, this structure marking the top of the flow. Above this is a bed of more compact lava than the lower. Crevices extending through both of these beds are filled with a calcareous segregation product.

I am unable to account for the occurrence of this immense lava formation, excepting by the supposition that the successive flows took place from an immense crack, the position of which is perhaps indicated by the great fault line along which the dislocation took place between the higher and lower plateau.

Terrace Deposit.[1]—The loam of this formation has been frequently mentioned in the previous pages. It occurs in the valley of every tributary of the Yang Ho and probably also of the Sankang Ho. It exists in the form of terraces between Chatau and Kiming, and these undoubtedly occur in the valley of the Sankang Ho from Paungan (chau) to Tatung (fu). Between the Kiming hill and the Papau mountain, a terrace of coarse detritus overlooks the valley of Hweilei (hien), its surface being several hundred feet above the Yang Ho.

In the valley of Siuenhwa (fu) this deposit seems to have suffered less from erosion, and rises, generally without terraces, at first gently then rapidly toward the bordering mountains, filling ravines high up their sides. Our road to the north lay over this deposit, as we skirted the hills between Siuenhwa and Kalgan, and we saw it fringing the Kalgan gorge with isolated terraces high above the river. Leaving this gorge, and ascending the valley of the Siwan creek, we found it in continuous terraces, which even at the Roman mission of Siwan, rise 200 or 300 feet above the creek.

Going southwest from Kalgan, we find this deposit continuous from the valley of Siuenhwa into that of Hwaingan, and we have already seen how it forms a plateau capping the ridge between this valley and the Yang Ho at Tienching. It is also undoubtedly represented along the Yang Ho from this place to Kalgan.

We have seen it, between Tienching and Yangkau, rising unbroken from the plain to high up the sides of the Barrier range, and continuous from here, in terraces, through the defile west of Yangkau into the valley of Kwantung (pu), and thence around the southern spur of the lower plateau through the valley of Fungching and the deep break in the higher plateau, west of Maanmiau, into the valley of the Té Hai, where its lofty terraces occupy the eastern part of this great depression.

The plain of the Kir Noor is formed by this deposit, which also extends through the valley on the east to the Si Ho tributary of the Yang Ho. As this formation

---

[1] For results, mostly negative, of a microscopical examination of the loam of this deposit from different localities, see Nos. 1, 2, 3, 5, 7, 8, 12, in Mr. Arthur Mead Edwards' Letter, Appendix 3.

is found at the head of the water system of this northern branch of the Yang Ho, it must be continuous, unless washed away, in all the valleys of this basin between the plateau and the Barrier range. Thus the deposit in the valley of the Kir Noor probably continues, through the break in the plateau to the southeast, into the valley of the Si Ho, and through this to the Yang Ho. Indeed, judging from the appearance of the region lying between the plateau and the Barrier range, as seen from the tower at Ha Noor, this deposit seems to occupy here a large area.

We can trace some of the more important islands that were isolated by the lake in which this deposit originated. One of these seems to have been that part of the plateau lying between the Si Ho and the Kir Noor. Another instance is the low ridge that separates the Yang Ho from the Hwaingan creek, while a much larger one is the hilly country between the Yang Ho and Sankang Ho.

Thus the body of water in which this deposit was formed consisted of a series of lakes several hundred feet deep, occupying the valleys of the Sankang Ho, Yang Ho, and Si Ho, and standing at a level sufficiently high to cover the lower watersheds between these streams.

This deposit is everywhere a calcareous loam formed of an almost impalpable powder, easily crushed between the fingers, and yet so firm that vertical cliffs of it remain unbroken for many years, which is sufficiently proved by the fact, before stated, that the inhabitants of the country excavate entire villages in the base of perpendicular cliffs that rise more than 100 feet above their dwellings. When breaks occur, the loam falls in immense plates, or tabular masses, leaving a new vertical face. Near the mountain sides and in the narrow gorges the loam is more sandy, and contains the gravel and fragments of rocks coming from the immediate neighborhood, but everywhere else it consists uniformly of an almost impalpable powder.

A characteristic feature of this loam deposit is its tendency to cleave according to two vertical planes at right angles to each other, causing it to assume the form of needles under certain conditions of erosion.

The effects of erosion in this deposit are often very interesting, illustrating in a marked manner the retrograde formation of ravines. The country is often cut up by gullies 30 to 70 feet deep, and from 10 to 20 feet wide, with vertical walls. In these channels wagon roads run for many miles without rising to the plain. In the valley, between Kwantung (pu) and the Yangkau defile, I crossed a gully 40 or 50 feet deep, and not more than four feet wide, having the same breadth all the way down, and which, with these dimensions, follows a tortuous course for more than a mile. In the same valley another ravine of this kind, only eight or nine feet wide, and not less than 100 feet deep, compelled us to make a detour of over a mile.

Wherever a cliff of this deposit presents itself the beginning of this action is visible. The surface drainage of a small neighboring area of the plain being concentrated toward one point on the edge of the cliff, cuts, in its fall, a channel from top to bottom, and this, with each succeeding rain, works its way backward toward the mountains. As the erosion progresses the sides of the gullies offer new starting points for tributary ravines.

We have here, in the softest material that can support such action, a repetition of the process which is causing the retrogression of Niagara falls, and which probably plays an important part in all valley erosion.

In intimate connection with this loam-deposit, stands the formation of the numerous isolated lakes met with on the route through the region we are now considering. I have frequently alluded to bars, or low watersheds, formed of the terrace-deposit, and stretching across valleys, causing the drainage to flow in opposite directions. These form the barriers to which almost every lake or pond, that has been mentioned, owed its existence after the retreat of the main body of the great inland sheet of fresh water.

We have seen that in those broad valleys where the lake-deposit has not been much subjected to erosion, its surface is not horizontal throughout, but rather, adapting itself to the general surface of the ground, or ancient valley, on which it lies, it rises from the centre to high on the sides of the surrounding mountains. Now when the sides of a valley approach each other and form a gorge connecting two broad enlargements of the valley, the terrace-deposit rises from the centres of both these basins, till it fills the gorge to about the same height as that at which it stands on the mountain sides around the basins. The height attained by the lake deposit in these narrow places is, in almost every instance, due to the fact that the usual deposit of loam was augmented by the large amount of detritus from the bordering hills.

As the large inland body of water disappeared and sank to the level of each of these bars, the sheet behind this remained isolated. In some instances the lakes thus formed have found outlets by cutting through their bars, but this was only where they received an important supply of water, derived from an extensive drainage area. In all other cases the barriers have suffered comparatively little from erosion.

Since their isolation these lakes have diminished in size, till they now possess but a small fraction of the volume necessary to fill their separate basins to a level with the surface of the inclosing bar.

I now propose to consider briefly the conclusions which the facts observed in this part of northern China seem to warrant.

The oldest stratified rocks seen throughout this region are highly metamorphosed and appear to belong to two distinct epochs; the hornblendic and chloritic series of schists representing the older, and the gneiss and granulite series, the younger.

After the deposition of the older metamorphic strata there seems to have been a disturbance producing folds with a trend between N. and W. Disturbances had also occurred by which the ridge between Nankau and Chatau was elevated and again depressed before the deposition of the great limestone formation, for the beds of this latter rest here immediately on the granite. Northwest of this ridge the limestone would seem to have been deposited in a shallower part of the sea, the character of the Hwaingan beds—which appear to represent the limestone—indicating the neighborhood of land.

After the deposition of the limestone strata these were traversed by the eruptive porphyries of Hiamaling, the debris of which form the chief ingredient of the conglomerate lying between the limestone and the coal-bearing series of Chaitang.

The next marked event was the forming of the coal-bearing rocks.

Although the disturbance, which was to produce the N. E. S. W. system of folds, appears to have been in operation before the deposition of the limestone, it was not until after the completion of the coal-bearing series, that this action cumulated in the great revolution by which the eastern portion of the continent received its outline, and the coal-bearing strata and older rocks were folded and prepared for the almost universal metamorphism that has affected them.[1]

An immense hiatus now occurs, for filling which there are no observed facts. This extends over the whole time that passed between the deposition of the coal-bearing rocks and the period of volcanic action in Southern Mongolia.

During this period occurred the eruption of the Kalgan trachytic porphyry and the deposition of its pluto-neptunian beds, and the outflowing on a gigantic scale, along the 41st parallel, of trachydoleritic and basaltic lavas.

The next phenomenon, of which the effects are visible, was the great dislocation by which at least the southern edge of the Mongolian plateau was raised. Near Fungching we have seen the high escarpment of the table-land, caused by this fault, trending away in a E. N. E. W. S. W. direction. If we produce this line toward the E. N. E. we shall find that it cuts the highest known point of the southern edge of the plateau—that near Ha Noor. The action of springs, that seem to rise along this fault line, is visible in the calcareous deposits seen near Maanmiau, and on the lower plateau near Fungching.

This great zone of volcanic action seems, as such, to mark the coast line of an extensive sea or ocean lying to the north, and it is an interesting fact that it lies nearly in a line with the axis of the Tienshan, in which we have every reason to believe that volcanoes still exist, though perhaps only as solfataras.

The dislocation by which the great escarpment of the plateau was formed, determined the depression between the table-land and the mountains south of it, which was to be occupied by the lakes already mentioned.

Before the deposition of the terrace deposit, the edge of the plateau had already been subjected to extensive erosion, by which great bays and channels were cut into it, and the valleys of the Té Hai and Kir Noor formed.

We come now to an interesting question—the origin of the chain of lakes so often referred to in the preceding pages, and of the deposit of loam by which they have recorded their former existence.[2]

That this deposit was formed in fresh water is shown by the presence of the shells found in the terrace of the Té Hai. The uniform character of the loam in the different basins, and in all parts of the same basin, its great extent, and the fineness of the material of which it consists, are conditions which prove that it is not of local origin, or derived from the detritus of the neighboring shores, but that it was brought into the lakes by one or more large rivers which must have drained an area of great extent. Now throughout the region in question, the only rivers are those of the Yang Ho and Sankang Ho basin, and, independently of the fact that these streams drain a very small area, the valley systems of these were almost entirely occupied by the lakes.

---

[1] See Chap. VII. [2] See Map XI, on Pl. 5.

Indeed the only direction from which a river of any importance could have come, was from the west, in which case it could only have been the Hwang Ho (Yellow river). Let us examine into the possibility of the existence of a communication between the valley of the Yellow river and the lake basins. When I was in the valley of the Té Hai, I saw distinctly that the break in the plateau continued to the W. S. W. as far as the eye could reach. A low, hilly country, much below the level of the plateau, appeared to shut in the valley at the distance of about twenty miles from the lake. Now on Klaproth's large map of Central Asia, on which, so far as my experience goes, the streams of this region are laid down with a remarkable approximation to accuracy, a branch of the Tourgen Gol[1] is given as rising in the very region occupied by the low hills observed by me. A native map of the province of Shansi, not always correct in its details, represents this stream as rising in the Té Hai.

Thus, I think, there is little doubt that a communication exists between the valley of the Té Hai and that of the Tourgen Gol, sufficiently depressed to be below the surface level of the terrace deposits. The Tourgen Gol is a tributary of the Yellow river, and if the watershed between the Té Hai and this river was below the level of the ancient lakes, these must have occupied part of the valley system of the north bend of the Yellow river, and must have left a corresponding deposit.

Now, although we have no information concerning the occurrence of the terrace deposit in the valley of the Tourgen Gol, we have direct testimony with regard to its existence over a large area in the land of the Ortous—the desert region inclosed by the northern bend of the Yellow river. Abbé Huc passed through this country on his way to Tibet, and describes it as a flat, sandy desert, frequently cut up by deep ravines, in the sides of which he observed, in one place, dwellings excavated in the same manner as those at Siwan.[2]

Indeed, all the information we possess concerning this region goes to show that it has been the basin of a great lake, which once extended from the northern bank of the Yellow river southwards to the mountains crowned by the Great Wall.[3]

Thus I think there can be little doubt that the terrace deposits, so common in the system of the Yang Ho, were precipitated in a chain of connected lakes, extending from Yenkingchau, N. N. W. of Peking, to near Ninghia (fu) in Kansuh, a

---

[1] Haishui of the Chinese. The valley of Tourgen Gol is probably also connected with the valley of the Kir Noor; see p. 29.

[2] "When the Chinese establish themselves in Tartary, if they find mountains the earth of which is hard and solid, they excavate caverns in their sides. These habitations are cheaper than houses, and less exposed to the irregularities of the seasons. They are generally well laid out; on each side of the door there are windows giving sufficient light to the interior; the walls, the ceiling, the furnaces, the kang, everything inside is coated with plaster so firm and shining that it has the appearance of stucco. These caves have the advantage of being warm in winter and cool in summer. . . . . . These dwellings were no novelty to us, for they abound in our mission of Siwan. However, we had never seen any so well constructed as these of the Ortous."—Abbé Huc, Travels in Tartary, etc., Vol. I, p. 180.

[3] Compare Ritter's Erdkunde. Asien, especially Vol. I, p. 153—160; also Huc, Vol. I, p. 235; and Travels of Gerbillon, in Du Halde.

distance of nearly 500 miles; and that this sediment was brought by the Yellow river and the tributaries of its upper course.

We have seen that the immediate cause of the formation of these lake basins is probably to be sought in the dislocation forming the plateau wall to the north of them, the descent of the land previous to that event having probably been toward the Gobi, in which direction also the Yellow river flowed, if it existed at that time.

The waters of the Yellow river filled the chain of basins thus inclosed between the plateau and the mountains forming the southern wall. There are now two channels by which the drainage of all this area finds its way to the Yellow sea, the Yang Ho gorge in the far east which opens on to the great plain west of Peking, and the deeply cut channel through which the Yellow river flows between Shansi and Shensi. Whether both of these outlets existed during the lake period, or only one of them, is a question of much interest in a physical-geographical point of view, for if all, or part, of the waters of the Yellow river flowed through the Yang Ho gorge, they found their way to the sea through the lower Pei Ho, a stream with which the Yellow river has united within historical times, after having flowed in an entirely different course, viz. its present one, in part, to the west and south of Shansi.[1]

The Yellow river flows, from Pauteh (chau) to the mouth of the Wei river, nearly 300 miles, almost due south, traversing, in deep gorges, two important mountain ranges which seem to be great anticlinal ridges of the limestone, and several minor ones. Considering these things, the regularity of its course is striking when compared with the winding courses common to rivers that cross parallel ranges, and the inclosed longitudinal valleys. The thought is suggested that the course of this channel may have been determined by a great crack.

In connection with this subject, I will add that it is certainly remarkable that the Chinese traditions of two great floods, often cited in the west, toward proving the universal belief in a general deluge, all point to this region. The earliest of these traditions is allegorical and goes back to a time, about 3100 B. C., when the yet barbarous founders of the nation were still living west of Shansi. "Kingkung fought with Chwanchio for the empire of the world; in his rage he struck, with his horn, the mountain Puchiau, which supports the pillars of heaven, and the bands of the earth were torn asunder. The heavens fell to the northwest, and the earth received a great crack in the southeast."[2]

The other tradition, preserved in the Shuking of Confucius, refers to a later date, and partakes of a more historical character. According to this account,[3] there was a great flood in the 61st year of the reign of Yao (2297 B. C.); the waters of the Yellow river mingling with those of the Yangtse Kiang, and threatening to overflow the mountains. A skilful engineer, Pekuen, worked nine years, without success,

---

[1] See Chap. V.

[2] Klaproth, Ritter's Asien, I, 158. Klaproth, in Asia Polyglotta, p. 28, comparing the dates of Hebrew, Brahminical, and Chinese traditions of deluges, obtains: Samaritan text, B. C. 3044, Brahminical date, B. C. 3101, Chinese, B. C. 3082.

[3] Ritter, Asien, I, p. 159. Compare Deguignes, Gesch. der Mongolen, Einleit. p. 4; and Mailla, Histoire générale de la Chine.

to effect a drainage; an object that was not accomplished until ten years afterward under the great Yu, by widening the channel of the river between Shansi and Shensi, especially in the gorges of Lungmun, Hukau, and Shanmun.

Mailla, one of the Jesuit missionaries employed in preparing the map of the empire, visited these localities, and relates that he saw with astonishment the remains of this gigantic enterprise.

However this may be, whether the works of Yu belong to the region of History or of Allegory, we have here two traditions, the first pointing to a convulsion causing a great flood, and perhaps also forming the channel between Shansi and Shensi; while the second evidently refers to an immense overflow of waters coming from the upper course of the Yellow river, and perhaps facilitated by obstructions in the narrow channel.

A gentleman, well versed in Chinese literature, informed me that, according to native authorities, the valley of the Yang Ho, between Chatau and Kiming, the easternmost of the ancient lake-basins, was once occupied by a lake which was drained, finally, by the Yang Ho gorge. Considering this, and the accounts of the Shuking, it is not, I think, impossible, that these traditions refer to the last events in the history of the lake period, and that within the memory of the Chinese people, a part at least of this great body of fresh water was still in existence, if, indeed, the formation of the channel between Shansi and Shensi, on which the retreat of the main body depended, does not also fall within this limit.

CHAPTER V.[1]

THE DELTA-PLAIN, AND THE HISTORICAL CHANGES IN THE COURSE OF THE YELLOW RIVER.

THE extent of the great plain of Eastern China is pretty well known from native and Jesuit authorities. It lies in a semicircle around the mountainous peninsula of Shantung. Its outer limit, as approximately given on the Jesuit map, begins in the department of Yungping (fu), and, running west, keeps south of the Great Wall till Changping (chau) N. W. of Peking. Thence, remaining east of the southern branch of the Great Wall, it follows a general S. S. W. course, passing westward of Chingting (fu) and Kwangping (fu), till it reaches the upper waters of the Wei river. Here it turns westward into Hwaiking (fu), and crosses the Yellow river in that department.

From the right bank of this river it trends a little east of south, passing west of Jüning (fu) (Honan), and then turning eastward it continues south of Kwang (chau) and north of Luhngan (chau) in Luchau (fu). Here an arm of the plain, in which lies the Tsau lake, stretches southward from the Hwai river to the Yangtse, and continues eastward on the right side of this river, occupying the region between the river and Hangchau bay. A hilly region, in the centre of which is Nanking, rises, like a large island from the plain, to the north of this arm.

The Shantung boundary of the plain begins at Laichau (fu), and after describing a great bow to the south it turns west at Shukwang (hien), and running thence to Changtsing (hien), in Tsinan (fu), it turns to the south and around to the southeast. Keeping this course it remains nearly parallel to the Imperial canal till the Kiangsu frontier, which it follows to the sea.

The greater part of the area included within these limits is a plain which seems to descend very gently toward the sea, and to be very generally below the high water level of the Hwang Ho. It is the delta of the Hwang Ho, and in part also of the Yangtse Kiang, and is remarkable for its semi-annular shape, half inclosing, as it does, the mountain-mass of Shantung.

The city of Peking stands on a raised border of loam, sand, clay, and gravel, which forms the northwestern skirt of the delta-lowlands, and seems to extend southward fringing the mountains along its western side. The name of the Talo lake (Ta great, and lo plateau or raised plain) seems to refer to such a border, and

---

[1] See Maps I—X, on Plates 4 and 5.

in the article on Kichau in the Yukung it is said that "the Lo (plateau) was drained."[1]

The fact, also, that in historical times none of the arms of the Hwang Ho have approached the western mountain border of the plain, both north and south of Kaifung, within a less distance than from ten to fifty miles, seems to point to the existence of a recent sea margin, which would be perhaps due rather to the detritus brought down by local streams than to the delta deposit of the Hwang Ho.

All the important changes in the lower course of the Hwang Ho have been recorded from early times by Chinese historians, and their documents and maps form the most complete history we possess of the wanderings of any river.

The Yukungchuchi (Peking, 1705), written by Chin Hu Wei, contains a series of maps in which these changes are laid down for a period of more than 3000 years. M. Biot has given the substance of that part of this work that relates to the Hwang Ho, in a carefully prepared paper.[2] I have, however, thought the subject to be one of sufficient interest to warrant the reproduction of the maps of Chin Hu Wei, with such explanations as will render them intelligible, without going beyond the limits of a work that is intended to give only my own contributions to the physiography of Eastern Asia. For farther information I must refer the reader to M. Biot's paper, of which I shall make use in explaining the maps.

In the Yukung, a chapter of the Shuking classic of Confucius, it is said that the course of the Hwang Ho was regulated by the Great Yu. Whether the works of Yu are to be understood as the labor of a single man, or as the results of the enterprise of a rising colony during several generations, there seems to be little doubt that more than 2000 years before the beginning of the Christian era the Chinese had brought this turbulent river under their control, by an immense system of dykes, and had begun to cultivate the extensive marshes of the delta plain.

Map No. 1 of the series, on plate 4, represents the course of the Hwang Ho as it existed, in the main, from the time of Yu down to 602 B. C.

Map No. 2 represents the course resulting from the first great change, that of the fifth year of the reign of Ting Wang (Chow dynasty), 602 B. C.

Map No. 3 serves to illustrate a passage in the writings of the poet Sse Ma Tsien, recording a diversion to the east and southeast. The easterly course, forming the Pien river, seems to have been the earliest recorded tendency of the river to follow its recent course. The opening of the first channels in this direction is given as occurring in 361 and 340 B. C.

The diversion, indicated on this map, through lake Yungtse to the southwest, happened, according to Sse Ma Tsien, towards the end of the Chow dynasty, during the third century before Christ.

Map No. 4 represents changes that occurred under Wutih (Han dynasty), about 132 B. C., when a great overflow toward the northeast took place, the river trending toward Kai (chau) in Chihli. At this time several arms were formed between

[1] E. Biot, Sur le chapitre Yukung, Journ. Asiatique, 1842.

[2] Sur les changements du cours inferieur du fleuve Jaune, Journ. Asiat. 1843.

Taming (fu) and the sea, which are also given. Previous to this, under Wentih, about 160 B. C., there was à breach formed at Yentsin near Kaifung.

Map No. 5 gives the second great change in the course of the "river of Yu," which occurred about 11 B. C., and was caused apparently by the blocking up of the channels leading to the Pei Ho.

Map No. 6 shows the channels as they existed during the Tang, and five succeeding dynasties, till the beginning of the Sung dynasty.

A note on the map of Chin Hu Wei says, "the course of the river remained the same from the time of Ming Ti (Tung Han dynasty) A. D. 70 . . . . . . till under Jin Tsung, A. D. 1034, when a break occurred at Hunglung, and another, fourteen years later, A. D. 1048, at Changwu, and the river of the Han and the Tang was entirely destroyed. The map covers a period of 977 years."

Map No. 7 (Pl. 5) represents the courses, under the Sung dynasty, from A. D. 1048 to A. D. 1194, a period of 146 years.

Map No. 8 records the course during the Kin dynasty. All the former channels appear blocked up, and the river, after entering Lake Lo, near the summit-level of the present Imperial canal, is seen to flow off to the N. E. through the Tatsing river, and to the S. E. through the Sz' river. Lake Lo appears from the observation of Clarke Abel, and from Chinese measurements, to be about 150 feet above the sea.

Map No. 9 shows the condition of the river under the Yuen and Ming dynasties, together with the Grand canal, a condition which seems to have remained substantially the same till within the last ten or fifteen years.

In early times the Yangtse entered the sea by three arms called the Sankiang, *i. e.*, "Three Rivers;" and Chin Hu Wei has given a map of these, founded on the opinions of early authorities. I have indicated them on map No. 1 of the series.

A glance at the nine maps of the delta courses will show how widely separated have been the limits of divergence of the arms of the Hwang Ho, within the past 3000 years. A mighty river, ever turbulent, subject yearly to an enormous increase in volume, an increase regulated rather by the amount of precipitation in the distant Kwenlun mountains, than by the local climate, it has ever been the terror of the countless millions through whose midst it flows.

From the earliest times an immense force has been at work to keep it from breaking through its dykes, or, when this has happened, to guide and retain it between new embankments. The quantity of solid material carried by the river and deposited along its course, is so great that its bed is rapidly raised, and appears to have been, before the last change, higher than the adjacent country.

Biot says, "it is certain that the bed of the river, from Hwaiking to the sea, is higher than the adjoining country."

Several times, during the great wars that have preceded the downfall of dynasties, this condition of the river has been turned to account as a weapon of offence. Breaking the embankments has been made to accomplish, almost instantaneously, by the destruction of hundreds of thousands of inhabitants, conquests that had been delayed by years of brave resistance.

From the earliest time of colonization on the delta-plain, the task of keeping the

Hwang Ho within its bed has been the constant care of the rulers of China, both when the country was united under one man, and when it has been subdivided into petty states. In the latter case in the treaties between states bordering on the Hwang Ho, the clauses regarding the regulation of that river appear to have been the most important and the most sacredly observed.

One of the most striking results of the official corruption that becomes general during the decay of a dynasty is the breaking loose of this great stream, as soon as the means for maintaining its embankments are misapplied.

The devastation caused by these overflows is awful beyond description. The loss of life is very great, and the destruction of the crops that form the means of support of millions, produces famine and the overrunning, by starving hordes, of the more fortunate districts of the adjacent country. The anarchy that rules in this struggle for life is almost beyond the conception of those who inhabit lands where the population is much below the capacity of the country, or which are easily reached by foreign supplies.

Within the last fifteen years one of these great changes has taken place, apparently from the same cause and with the same effect as above indicated. Instead of emptying into the Hwang Hai, or Yellow Sea, the Hwang Ho now has its mouth in the Gulf of Pechele, which it enters through the Tatsing river. The old mouth of the river was found to be dry in 1858.

According to information furnished to the Rev. Mr. Edkins, by officials of the Board of Foreign Affairs at Peking, the principal break occurred at Fungpeh (ting) in Süchau (fu), the waters flowing away to the N. E. In Tsinan (fu), the capital of Shantung, the waters of the Tatsing river are increased to six times their original volume by the contributions of the Hwang Ho.

In 1863 the river had not yet determined a channel, but its waters were spread over large tracts of country, and the city of Wuting (fu), nearly sixty miles north of Tsinan (fu), was almost inaccessible.

The present course of the Hwang Ho is indicated, so far as known, on Map No. 10.

Owing to the great quantity of material brought down by this river, and to the absence of great oceanic currents, that might, if present, interfere with its deposition, the delta is rapidly increasing in size, and the adjoining seas are becoming shallower.[1]

Probably nowhere can the rate of growth of deltas be better studied than in China. Cities that were built on the delta plain of the Hwang Ho several thousand years since are still in existence, together with the archives of their history. In the cases of those that were built near the sea, the distances from this are given; and frequent mention is made of towns, mounds, and natural hills, washed by the sea, within historical times, which are now far inland.

Thus, in B. C. 220, the town Putai is said to have been 1 li west of the sea-shore, while in A.D. 1730 it was 140 li inland,[2] a yearly increase of 100 feet, more or less,

---

[1] Barrow estimated the hourly discharge of sediment at 2,000,000 cubic feet.

[2] Fangyuchiyau; Chihli.

according to the length of the li. Hienshuikau (on the Pei Ho, in long. 117° 32′ E.) is said to have been on the sea-shore in A. D. 500,[1] and is at present about eighteen miles distant, an increase of about 81 feet per annum.

Along the southern shore of the gulf of Pechele the yearly increase N. E. of Shukwang since B. C. 220, seems to have been not more than 30 feet.

The sea-shore, according to local tradition, was near the present location of Tientsin (fu) during the Han dynasty.

It is also recorded that under the reign of the Han, the Hwang Ho entered the sea at Changwu, near the present Tsinghai.[2]

---

[1] Fangyuchiyau; Chihli.

[2] Ibid.

CHAPTER VI.[1]

ON THE GENERAL GEOLOGY OF CHINA PROPER; A GENERALIZATION BASED ON OBSERVATIONS, AND ON THE MINERAL PRODUCTIONS AND THE CONFIGURATION OF THE SURFACE.

It is with much misgiving that I begin even an attempt at a general sketch of the geology of China. The great extent of the country, the very limited area examined geologically, the, mostly, very general character of the observations made within that area, and our ignorance of the geological structure of the surrounding countries, render the attempt more than dangerous.

The sketch, and the map accompanying it, make no claims to accuracy, but I hope to show by means of them the leading features of the structure of the country, as deduced from observations in parts of the country and from mineral productions. The fact that hardly any two maps of China resemble each other in the geographical names; and that on most of them many of the names that I must use are not given, renders a sketch-map necessary, and this is to be regarded as a colored guide to the generalizations, and not as a geological map of the country.

The data on which the generalizations are founded consist in:—

My own observations.

The observations of other European travellers.

And in the information obtained from Chinese authorities.

The limits of my own observations have been already given; they were confined to the valley of the Yangtse Kiang, from the sea to near the eastern boundary of Sz'chuen, and to the northern departments of the provinces of Chihli and Shansi. The results of this portion of the data have been given in the preceding pages.

The observations of European travellers have furnished, so far as my knowledge of them goes, but very little information on the geology of the country, and even this is often vague and evidently incorrect. I have thought it worth while to give, in a condensed form, such information as I have been able to extract from this source.

Nanking to Canton.[2]—Gray, compact limestone is quarried back of Nanking. Siaukushan [Little Orphan Island], near the mouth of Poyang lake, is pudding-

---

[1] See Map, Pl. 6.

[2] Clarke Abel. Narrative of a Journey in the Interior of China, and of a Voyage to and from the Country, 1816—1817, etc. Lond. 1818.

stone (?). The high Liushan [west of Poyang lake and south of Kiukiang] are of fine-grained granite and micaceous schist poor in quartz, in vertical strata trending N. E. S. W.[1] On the left bank of the Kan river, above Kihngan (fu), there is sandstone. Between Wanngan (hien) and Kanchau (fu) there is dark gray schist resting on granite. Black slate occurs between Kanchau (fu) and Nanngan (fu). The summit of the Meiling pass is of argillaceous sandstone, immediately south of which begins limestone. Between Nanhiung (fu) and Shauchau (fu) the limestone ceases and is followed by red sandstone with coal seams. Nearer to Shauchau (fu) there is limestone resting on a breccia of limestone, calcareous red sandstone, and quartz, the whole cemented by limestone. Near Yingting (hien) there is grayish-black limestone in which is the cavern of Kwangsin. Hills of grayish-yellow, argillaceous sandstone, with veins of quartz, occur about half way between Yingting (hien) and Hingyuen (hien); [on Abel's route map the whole country between these two places is represented as sandstone.] The coal brought to Abel from the towns on the Yangtse resembled cannel coal, that in Kiangsi "bovey" coal.

At Fuhutang (on the Kan river), soon after leaving the Poyang lake, there are vertical coal pits. The fragments at the bottom of the hill where these are situated appeared to be pure slate.[2]

Canton to Hankau through Hunan.[3]—The rocks noticed on the North river (Peh kiang) were red sandstone and limestone. Four miles inland from Pangkwang there are coal mines, belonging to the government, 40 to 50 feet deep. Red sandstone occurs along the boundary between Kwangtung and Hunan on the Meiling pass. Red sandstone occurs near Shachulung, a coal village on the north slope of the Nanling near the end of the Meiling pass. A few miles below Laiyang (hien) there are limestone quarries. At Pingtan, a few miles below Siangtan (hien), there are limekilns and quarries of limestone. Sandstone is quarried at Kingtsewan, about twelve miles below Changsha (fu).

Chehkiang and Fuhkien.[4]—About ten to fifteen miles west of Yenchau (fu) (Chehkiang) are limestone mountains, and a few miles farther west beautiful green granite. Near Hwuichau (fu) (Nganhwui) the hills consist of a red sandstone resting on slate. Near Küchau (fu) (Chehkiang) there is red, calcareous sandstone. The road on the pass between the Shangyang river and the Chehkiang river is paved with granite. The road at the N. W. foot of the Bohea mountains leading from Hokau, in Kwangsin (fu) (Kiangsi), into Fuhkien, is paved with granite. The rocks at Wuishan, on the east side of the Bohea mountains in Fuhkien "consist of clay slate, in which occur, embedded in the form of beds or dykes, quartz rock, while granite of a deep black color, owing to the mica which is of a fine deep bluish black, cuts through them in all directions." "Resting on this clay slate are sandstone conglomerates formed principally of angular masses of quartz, held together by a calcareous basis, and alternating with these cönglomerates there is a fine, calcareous,

---

[1] Ritter, Asien, III, p. 675, citing Ellis' Journal, p. 342, and Clarke Abel, p. 167.

[2] Ellis' Journal, II, p. 107.

[3] Rev. Mr. Bonny. A Trip from Canton to Shanghai. Pamphlet. Shanghai, 1861.

[4] Fortune. Tea Districts, etc.

granular sandstone in which beds of dolomitic limestone occur." "Granite forms the summits of most of the principal mountains in this part of the country."

Canton to the Sea.[1]—A gray-wacke, containing much quartz, forms the hills near Canton. Underneath this rock is red sandstone, "varying from a bright red, fine-grained rock to a coarse conglomerate, full of large pebbles of quartz." These strata dip to westward. Granite occurs below the sandstone and crops out more and more, as the river approaches the sea. Near the coast the granite forms peaks 1,200 to 2,000 feet high, which continue as barren islets toward the island of Hainan.

Kingyuen (fu) in Kwangsi.[2]—The marble mountains south of Kingyuen (fu) give rise to innumerable large springs, and even rivers disappear in them to come again to light after following long subterranean courses. The many colored varieties of marble of this region are celebrated, and the marble formation (Marmor Gebirge) seems to predominate.

Salt Wells of Sz'chuen.[3]—M. Imbert has given a vivid description of these, and although it has often been quoted, it is sufficiently interesting to be inserted here.[4] These are at Wutung, in the department of Kiating (fu), and near the city Kiating.

"There are some ten thousand of these springs, or artificial brinepits, in a space about ten leagues long and four or five leagues broad. The Chinese effect the boring of these pits with time and extreme patience; yet with less expense than with us. They have not the art of working rocks by mining (blasting?); yet all the pits are constructed in the rock. These pits are commonly from 1,500 to 1,800 feet (French) deep, and are only five or at the most six inches in diameter. These little wells, or tubes, are perpendicular, and as polished as glass. Sometimes the entire depth is not continued in solid rock, but the workmen encounter beds of shale, coal, etc.; then the operation becomes more difficult, and sometimes fruitless; for as these substances do not offer a uniform resistance, it sometimes occurs that the shafts lose their perpendicularity; but these are rare cases. When the rock is favorable, they advance at the rate of two feet in the twenty-four hours. It requires at least three years to sink one pit." A pit of this kind costs about 1,000 taels of silver.[5] "The mode of pumping is exceedingly simple, yet laborious; being effected chiefly by manual labor. The water is very briny, giving, by evaporation, a fifth or more, and sometimes one-fourth, of salt."

"The air, which escapes from these pits, is very inflammable. If a torch is presented to the mouth of the shaft, the gas ignites, with a great column of fire, from twenty to thirty feet in height, exploding with the rapidity of powder." This gas is conducted through bamboo tubes to the saltpans under which it is burned to effect the evaporation. "Sometimes, in boring the salt pits, very thick beds of coal are passed through at a depth of several hundred feet." "In sinking these wells a bituminous oil [petroleum], which burns in water, is commonly found at a depth of about 1,000 feet. They collect daily four or five jars of 100 pounds each. This

---

[1] Chinese Repository, III, p. 87.

[2] Ritter, Asien, III, 758.

[3] Imbert, Annales de l'association pour la propagation de la foi. Vol. III, p. 369.

[4] The extract given here is taken from R. C. Taylor, Statistics of Coal, Phil. 1848, p. 660, with some remarks from Chinese Repository, XIX, p. 325.

[5] 1 Tael = $1.33.

oil has a very powerful odor, and is used to light the area where the pits and coppers of salt are concentrated."

"The largest fire wells are those at Tsélieoutsing, forty leagues from Wutung. Tsélieoutsing, situated in the mountains, on the banks of a small river, also contains salt pits, bored in the same manner as at Wutung. In one valley are seen four pits which give a flame, to an amount truly frightful, but no water. These pits, for the most part, have previously afforded salt water; which water being drained, the proprietors, twelve years since, caused them to be sunk even to *three thousand feet and more* of depth, hoping to procure an abundant supply of water. All this was in vain; but there suddenly gushed forth an enormous column of air which brought with it large, dark particles. These did not resemble smoke, but the vapor of a glowing furnace. This air escaped with a roaring and frightful rumbling, which was heard at a great distance. The orifices of the pits are surmounted by a wall of stone six or seven feet high, for fear that, inadvertently, or through malice, some one might apply fire to the opening of the shaft. This misfortune happened in August last. As soon as the fire was applied to the surface of the well, it made a frightful explosion, and even something was felt approaching to an earthquake. The flame, which was about two feet high, leaped over the surface of the earth without burning anything. Four men devoted themselves and carried an enormous stone over the orifice of the pit. Immediately it was thrown up into the air; three of the men were scorched, the fourth escaped; neither water nor dirt would extinguish the fire. Finally, after fifteen days of stubborn work, a quantity of water was brought over the neighboring mountain, a lake or dam was formed, and the water was suddenly let loose, which extinguished the fire. This was at an expense of about thirty thousand francs."[1]

Fossils from China.[2]—Mr. Davidson, after examining a collection of shells sent by Dr. Lockhart to the British Museum, came to the conclusion, "that the specimens belonged to eight Devonian species, seven of which are common to several European localities, among which we may mention Ferques and Néhon (France), Belgium, and the Eifel, but they are not found all existing together in any one of these localities. In external aspect they most resemble those from Ferques, in which locality, however, neither the *Cyrtia Murchisoniana* nor the *Rhynchonella Hanburii* have been as yet discovered." If to these we add the other two described by M. de Koninck,[3] the total number of Chinese Devonian types now known will amount to ten species: viz., 3 of Spirifer, 2 of Rhynchonella, 1 Productus, 1 Crania, 1 Cornulites, 1 Spirorbis, and 1 Aulopora. The species determined by Mr. Davidson were as follows: Spirifer disjunctus, *Sowerby;* Cyrtia Murchisoniana, *De Koninck;* Rhynchonella Hanburii, *Davidson;* Productus subaculeatus, *Murchison;* Crania obsoleta, *Goldfuss;* Spirorbis omphalodes, *Goldfuss* (?); Cornulites epithonia, *Goldfuss* (?);

[1] Compare Humboldt, Asie Centrale, II, p. 521, 525.

[2] On some Fossil Brachiopodes, etc. T. Davidson. Quart. Journ. Geolog. Soc., IX, 1853, p. 353.

[3] "Notice sur deux espèces Brachiopodes du Terrain Paléozoique de la Chine." Bullétin de l'Académie Roy. des Sciences, Lettres et Beaux Arts de Belgique. 1846. XIII, pt. 2, p. 415.

Aulopora tubaeformis, *Goldfuss;* Spirifer Chechiel, *De Koninck;* Rhynchonella Yuenamensis, *De Koninck.*

Some fossil brachiopods from Gouchouc, twenty leagues W. S. W. from Patang on the Kinsha Kiang, and near the Tibet-Sz'chuen frontier, were determined by M. Guyerdet[1] as follows: Terebratula cuboides, *Sow*, carb. and Devon., figured in Descript. des Anim. foss. de la Belgique, De Koninck, 1842—1844, p. 285. Terebratula reticularis, *Linné*, Devonian; figured in Russia and the Ural mountains: Murchison and v. Keyserling, II, 90. Terebratula pugnus, *Martin;* figured in Sowerby, Conchyl. pl. ccccxcvii. Mr. Woodward has described an Orthoceras from China.[2]

Hoshan (*Fire Mountains*).—These are without doubt burning seams of coal. One of these burning mountains, called Hoyau, occurs 55 li N. W. of Kwangling in Tatung (fu), Shansi.[3]

Sir R. I. Murchison speaks of some Upper Devonian fossils, from Sz'chuen, given to him by Dr. W. Lockhart, as "identical in specific character with Spirifer Verneuilii, S. Archiaci, Productus subaculeatus, and other European forms."[4]

I was told by the Rev. Mr. Edkins that the island of Situngting in the Taihu lake (west of Shanghai) contains fossiliferous limestone.

In the following table are given a large number of localities of coal and alum (the latter is made in China, I believe, always from pyritiferous shales that accompany coal), to be used in locating the coal-bearing formation; and of indications of limestone, as limestone-marbles, limestone, caves, stalactites, fossil brachiopods, etc. These localities are in every instance, unless otherwise stated, taken from Chinese geographical works, especially from the Tatsingytungchi, and the geographies of the separate provinces.

This is followed by a table of salt wells in Yunnan and Sz'chuen, which will be explained further on; and by a table of gold-bearing localities to assist in locating the granito-metamorphic formation.

---

1 Comtes Rendus. Acad. des Sciences, Paris, 1864, LVIII, No. 19, p. 878.

2 Quart. Journ. Geol. Soc., 1856, p. 379.

3 Biot, in Journ. Asiat., 1840, October.

4 Siluria, p. 425. Lond. 1859.

TABLE OF LOCALITIES OF COAL, ALUM, LIMESTONE, LIMESTONE-MARBLES, FOSSILS, CAVES, STALACTITES, ETC., IN CHINA.[1]

F = fu; C = chau; H = hien.

Province.	Department.	District.	Place and circumstances of occurrence.
Chihli.	Shuntien F.	Fangshan H.	Anthracite, S. W. 40 li at Hwanglung Mt., white marble.
	" "	Wangping H.	Anthracite at Muntakau, Maanshan, and Tatsau.
	" "	" "	Bituminous at Chaitang and Chingshui.
	" "	Waitso H.	White marble.
	Yungping F.	Funing H.	70 li N. E. at Liulu Mt., coal. At Shïling, coal.
	Kwangping F.	Tsz C.	Coal.
	Siuenhwa F.	- - - - -	Coal at Kingtingpu.
	" "	Yü C.	Anthracite (Shïtan).
	" "	Paungan C.	Anthracite (Shïtan). Coal in hills north of Sin-paungan.
	" "	" "	Anthracite at Kiming.
	" "	Sining H.	Anthracite (Shïtan).
	" "	Wantsuen H.	Anthracite (Shïtan). 15 li S. brown coal at Wu-taiyau.
	" "	- - - - -	Brown coal 60 li N. N. W. of Kalgañ at Wushïkia.
	" "	- - - - -	Coal at Siautungko 180 li W. of Kalgan.
	Pauting F.	Y. C.	Great cavern in Mt. Lungchi. (B.)
	Chingting F.	- - - - -	Several large caverns.
	Shunteh F.	- - - - -	Several large caverns.
Shansi.	Taiyuen F.	Chauyang H.	Large caverns near Chauyang H. 100 li E. of Taiyuen F. (B.)
	" "	- - - - -	Large caverns near Tseuhong. (B.)
	" "	- - - - -	Coal 12 miles S. W. of Taiyuen on W. side of Făn R. (Bagl.)
	" "	- - - - -	Coal 35 miles S. W. of Taiyuen on W. side of Făn R. (Bagl.)
	" "	- - - - -	Lime burnt, 30 miles S. W. of Taiyuen on W. side of Făn R (Bagl.)
	Pingting C.	Soyang H.	Anthracite (Shïtan).—Alum.
	" "	- - - - -	Coal 12 W. of Pingting C. (Bagl.)
	Hin C.	Tsingloh H.	Anthracite (Shïtan).
	Tatung F.	- - - - -	Bituminous coal "quarried" in large blocks (Ta-tan) near the city.
	" "	Kwangling H.	Coal.
	" "	Lingkiu H.	Stalactites in Mt. Peshan.
	Fănchau F.	- - - - -	Coal and lime 17 miles S. of city in the range east of Făn R. (Bagl.)
	" "	Ling H.	Coal 70 li E.
	Pingyang F.	Yching H.	Anthracite (Shïtan).
	" "	Yoyang H.	Anthracite (Shïtan).
	" "	Lingfung H.	Anthracite (Shïtan) near Pingyang.
	" "	Hungtung H.	Anthracite (Shïtan).
	" "	Fehshan H.	Anthracite (Shïtan).
	" "	Taning H.	Great caverns 20 li N. W. in Mt. Kung.
	" "	Kih C.	Lime.—Alum.
	Hoh C.	Lingshï H.	Anthracite (Shïtan).
	Tsehchau F.	Yangching H.	Anthracite (Shïtan).
	Kiang C.	Yuenchü H.	Alum.
	Kiai C.	- - - - -	Alum.
	" "	Ngany H.	"Cave of the Winds" S
Shensi.	Yulin F.	Yulin H.	Anthracite (Shïtan) 20 li S. E. at Mt. Tan.
	Tungchau F.	Chingching H.	Alum.
	" "	Tungkwei H.	Alum.
	Fungtsiang F.	Kienyang H.	Cavern, 30 li S. E.
	Ningkiang C.	- - - - -	Fossil Brachiopods (Shïyen).

[1] B. = Biot; Bagl. = Rev. P. Bagley; Edk. = Rev. Mr. Edkins.

TABLE OF LOCALITIES OF COAL, ALUM, LIMESTONE, LIMESTONE-MARBLES, &c.—*Continued.*

Province.	Department.	District.	Place and circumstances of occurrence.
Shensi.	Hanchung F.	- - - - -	Fossil Brachiopods (Shïyen).—Many large caverns. (B.)
	Yenngan F.	Yenchuen H.	Petroleum springs.
	Tungchau F.	- - - - -	Coal 15 miles above junction of Făn R. and Hwang Ho. (Bagl.) Many caverns in the Tsepe, Lungmun, Taney, and Seou mountains. (B.)
Kansuh.	Lanchau F.	- - - - -	Coal 40 li S. W.
	" "	Titau C.	Coal 80 li distant.
	" "	Kin H.	Coal 40 li N. W.
	Kungchang F.	Tungwei H.	Coal 60 li S. E. at Lieutungping.
	Tsin C.	Tsinngan H.	Coal 10 li N. W. at Sulungpu.
	Ninghia F.	- - - - -	Coal N. E. on opposite bank of Hwang Ho (Huc).
	Liangchau F.	Yungchang H.	Anthracite (Shïtan) 20 li S. E. at Mt. Tan.
Jehho.	Chingteh F.	- - - - -	"Bad coal" 40 li S. E. at Mangninchuenkau.
	" "	- - - - -	Anthracite, E. near Sankia, W. of Palisade.
	" "	- - - - -	Anthracite and bituminous coal 40 li E. of Sankia.
	" "	- - - - -	Much coal among the mountains along the Palisade.
Shingking.	- - - - -	Kaiping H.	Anthracite.
	- - - - -	- - - - -	Coal on W. coast of Liautung promontory in lat. 39° 40′.
	- - - - -	- - - - -	Coal S. E. of mouth of Liau R.
	- - - - -	Chauyang H.	Coal at Latsz Mt.
Shantung.	Tsingchau F.	Yihte H.	Coal and alum at Yehchintsung.
	Taingan F.	- - - - -	Stalactites.
	Ichau F.	Kü C.	Stalactites, 150 li N. at Yünkungshan.
	Tsinan F.	- - - - -	Much coal in the range, 33 miles E. (Bagl.)
Kiangsuh.	Kiangning F.	- - - - -	Coal at Chunhwachen half-way between Kinyang H. and Nanking. (Edk.)
	" "	Kiangpu H.	Great cave ("Pit of Heaven") 30 li W.
	Chinkiang F.	Kintang H.	Stalactites 65 li W. at Mt. Mau.
	Süchau F.	Siau H.	Anthracite and lime 30 li S. E. at Peitutsung on Mt. Peitu.
	Süchau F.	- - - - -	Marble on islands of Taihu lake.
Nganhwui.	Ningkwoh F.	In all the H.	Anthracite (Shïtan).
	Taiping F.	Fanchang H.	Brown coal? (Kaufung.)
	Ho C.	Heishan H.	Coal.
	Luchau F.	Tsau H.	Large cavern near town.
	" "	Luhkiang H.	Alum.
	Fungyang F.	- - - - -	Alum.
Honan.	Honan F.	Kung H.	Coal.
	" "	Loyang H.	Coal.
	" "	Tungfung H.	Stalactites in Mt. Sansz.
	Ju C.	Lusan H.	Coal.
Hupeh.	Ichang F.	Kwei C.	Coal on banks of Yangtsekiang.
	" "	Patung H.	Coal on banks of Yangtsekiang.
	Yunyang F.	Fang H.	Stalactites.—Alum.
	Kingchau F.	Changyang H.	Cavern in Mt. Fang.
Sz'chuen.	Süchau F.	- - - - -	Coal on Yangtsekiang near the city. Coal at Lotu.
	Kiating F.	Kienwei H.	Coal in the salt district. (Imbert.)
	" "	- - - - -	24 caves in a mountain near the salt wells.
	Chungking F.	- - - - -	Coal.
	Chung C.	- - - - -	White marble 70 li N. W. at Mt. Peishï.
	Tungchuen F.	Pungchi H.	Limestone 90 li S. E.
Chehkiang.	Hangchau F.	In all the H.	Limestone in all the mountains of the department.
	" "	- - - - -	Many caverns in Mt. Pelaifung.
	" "	Changhwa H.	Fossil Brachiopods (Shïyen) in Shïyen cave at Mt. Yunko.
	Huchau F.	- - - - -	Coal.—Stalactites in Wanglung cavern.
	Wanchau F.	Pingyang H.	Alum.

TABLE OF LOCALITIES OF COAL, ALUM, LIMESTONE, LIMESTONE-MARBLES, &c.—*Continued.*

Province.	Department.	District.	Place and circumstances of occurrence.
Chehkiang.	Chuchau F.	- - - - -	Caverns in many of the mountains.
	" "	Lungtsiuen H.	Fossil Brachiopods and a cavern on Mt. Wang-matsien.
	Shauhing F.	- - - - -	Caverns.
	Taichau F.	- - - - -	White marble on Mt. Tsang.
	Kinhwa F.	Kinhwa H.	Cavern (Tsutsesantung).
	" "	Lanki H.	White stalactites at Peiyün cave in Mt. Tungnien.
	" "	" "	Lime at Peikang Mt.
	Yenchau F.	Tsenngan H.	Stalactites.
	" "	Tunglu H.	Stalactites at Langsien cave.
	" "	Fănshui H.	Cavern (Yangsantung).
	Küchau F.	Singan H.	Coal.
	" "	Kiangshan H.	Coal (Chin. Rep. xix, 387).
	" "	Changshan H.	Coal (Chin. Rep. xix, 387).
Kiangsi.	Nanchang F.	Fungsin H.	Anthracite at Lauhukau.
	Yuenchau F.	Pinghiang H.	Anthracite.
	" "	Făni H.	Cavern and Fossil Brachiopods.
	" "	Wantsui H.	Fossil Brachiopods.
	Kwangsin F.	- - - - -	Coal (Chin. Rep. xix, 387).
	" "	Tsienshan H.	Alum.
	Linkiang F.	Sinyü H.	Stalactites.
Hunan.	Changsha F.	Liuyang H.	Alum.
	Hăngchau F.	Hăngshan H.	Coal. Fossil Brachiopods at Mt. Nesho.
	" "	Laiyang H.	Coal.
	" "	In all the H.	Alum.
	Pauking F.	Siying H.	Coal.
	Kweiyang C.	- - - - -	Fossil Brachiopods at Mt. Shïyen.
	" "	In all the H.	Alum.
	Yungchau F.	Lingling H.	Fossil Brachiopods.
	Changteh F.	Nganhiang H.	Fossil Brachiopods.
Kweichau.	Chinyuen F.	- - - - -	White marble just east of the city.
	Shihtsien F.	- - - - -	"Dragon Cavern" 1 mile S. W. of city.
Yunnan.	Wuting C.	Yuenmau H.	Alum. Caves with bones. Fossil Brachiopods in the Kauhyin Mt.
	Yungchang F.	- - - - -	Caverns.
	Yauking? F.	- - - - -	Caverns. (B.)
	Tali F.	- - - - -	Orthoceratites.
Fuhkien.	Hinghwa F.	- - - - -	Coal (Chin. Rep. xvi, p. 80).
	- - - - -	Anko	Anthracite (Chin. Rep. xvi, p. 80).
	Changchau F.	- - - - -	Caverns.
	Funing F.	- - - - -	Caverns.
	Tsiuenchau F.	- - - - -	Caverns.
Kwangtung.	Shauchau F.	- - - - -	Coal.
	" "	Juyuen H.	Stalactites.
	Shauking F.	- - - - -	Stalactites and Fossil Brachiopods at Mt. Shï-yen.
	" "	- - - - -	Dendritic marble.
	Lienchau F.	- - - - -	Stalactites.
Kwangsi.	Kingyuen F.	- - - - -	Ossiferous caverns in the Nanshan Mts.
	Kweilin F.	- - - - -	Fossil Brachiopods.—Stalactites.
	Pingloh F.	Pingloh H.	Stalactites 31 li E.
	" "	Kungching H.	Stalactites 5 li E. at Mt. Kintsumi, and 28 li E. at Mt. Yintieh.
	" "	Lipu H.	Stalactites 1 li S. at Mt. Sung.
	Wuchau F.	Tsinki H.	White marble 10 li N. at Peishï.
	" "	Hwaitsih H.	Marble 80 li S. W.
	Yulin C.	Pohpeh H.	Stalactites 30 li S.
	Sinchau F.	Pingnan H.	Fossil Brachiopods 12 li S. E. at Mt. Yenshï.
	Nanning F.	Suenhwa H.	Fossil Brachiopods 90 li E. at Mt. Shïyen.
	Taiping F.	Shangsz C.	Stalactites and white marble 2 li E. at Mt. Peishï.

TABLE OF LOCALITIES PRODUCING SALT FROM ARTESIAN WELLS.

Province.	Department.	District.	Place and circumstances of occurrence.
Sz'chuen.	Chingtu F.	- - - - -	Wells.
	" "	Kien C.	Wells.
	Tsz C.	- - - - -	80 wells.
	" "	Tszyang H.	4 wells.
	" "	Nekiang H.	2 wells.
	" "	Jinshan H.	10 wells.
	" "	Tsingnien H.	237 wells.
	Ningyuen F.	Hwuili C.	Wells.
	" "	Yenyuen H.	Wells.
	Pauning F.	Langtsung H.	Wells.
	" "	Naupu H.	Wells.
	Shunking F.	In all the H.	Wells.
	Süchau F.	Fushun H.	Wells.
	Chungking F.	Pah H.	Wells.
	" "	Pihshau H.	Wells.
	Chung C.	- - - - -	Wells.
	Kweichau F.	Wan H.	Wells.
	" "	Wushan H.	Wells.
	" "	Yunyang H.	Wells.
	" "	Fungtsi H.	Wells.
	" "	Kai H.	Wells.
	Suiting F.	Tatsoh H.	Wells.
	Tungchuen F.	In all the H.	Wells.
	Mei C.	Pangshan H.	Wells.
	Kiating F.	Weiyuen H.	Wells.
	" "	Yung H.	Wells.
	" "	Tiewei H.	Wells.
	" "	Lohshan H.	Wells.
	Kung C.	Puhkiang H.	Wells.
	Lu C.	Kiangngan H.	11 wells N. W. of town
Yunnan.	Yunnan F.	Nganning C.	80 wells.
	Tali F.	Yunglung C.	Wells.
	" "	Langkiung H.	Wells.
	Tsuhiung F.	Tingyuen H.	Wells of black salt.
	" "	Kwantung H.	Wells of black salt.
	" "	Yau C.	Wells.
	Wuting C.	Tsauchitsing.	Wells.
	" "	Yuenmo H.	Well at Tsukiutsing.
	Likiang F.	- - - - -	Wells at Sipeh Mt.
	Pu'rh F.	Ningurh H.	Red salt.
	Kingtung (Ting)	- - - - -	Wells.
	Yungpeh (Ting)	- - - - -	Wells.
Shensi.	Kia C.	- - - - -	Lake salt.
	Yulin F.	Yulin H.	Lake salt 80 li S. at Yühopu.
	" "	Tingpien H.	Salt lake N. W. at Yentsangpu ("salt mine").
Shansi.	Taiyuen F.	Taiyuen H.	Salt.
	" "	Tsingyuen H.	Salt.
	Hin C.	Tingsiang H.	Salt.
	Kiai C.	Ngani H.	Salt lake.
	Tatung F.	Tatung H.	Salt.
	" "	Hwanyuen C.	Salt.
	" "	Ying C.	"Excellent salt at Yanghochiao."
	Lungan F.	- - - - -	Salt.
	Pauteh C.	- - - - -	Salt.
	Hoh C.	- - - - -	Salt.
	Sieh C.	- - - - -	Salt.

TABLE OF GOLD WASHINGS AND MINES.

Province.	Department.	District.	Place and circumstances of occurrence.
Chihli.	Shuntien F.[1]	Miyun H.	Gold mine 8 li E. of city.
	Yungping F.	Tsienngan H.	Gold washings in the Kwaihochuen R.
	" "	Lulung H.	On Mt. Tsu.
Shensi.	Singan F.	Lintung H.	On Li Mt. 2 li W. of city.
	Shang C.	Lohngan H.	Coarse wash gold at Hwanglungshan 80 li N. E. of city; and rich washings at Yanghwashan.
	Hanchung F.	Sihiang H.	Gold.
	Hinngan F.	Hanying (ting)	Coarse gold in the Han R.
Kansuh.	Lanchau F.	- - - - - -	Coarse wash gold.
	Kungchang F.	Min C.	Coarse wash gold.
	Kiai C.	Wăn H.	Coarse wash gold.
	Sining F.	Sining H.	Coarse wash gold.
	Suh C.	- - - - - -	Gold 70 li W. of the city at Tungtingshan.
	Chinsi.[2]	- - - - - -	Gold 60 li E. at Kinshan.
Shantung.	Ichau F.	Lanshan H.	Gold and silver mine 90 li S. W. at Paushan, and gold 60 li N.
	" "	Kü C.	Gold 100 li N. at Chipaushan.
	Tsingchau F.	Linkü H.	Gold-sand 60 li S. W. at Sungshan.
	Tungchau F.	- - - - - -	Gold.
Hupeh.	Hwangchau F.	Hwangkang H.	Wash gold 140 li N. at Tankingshan.
	" "	Hwangan H.	Gold E. at Tsangkiashan.
	Kingchau F.	- - - - - -	Gold.
	Shinan F.	Kienchi H.	Coarse wash gold 15 li W. at Shïjoushan.
Sz'chuen.	Chingtu F.	Kien C.	Coarse wash gold.
	" "	Wangkiang H.	Coarse wash gold.
	" "	Tsungking H.	Coarse wash gold.
	" "	Pang H.	Coarse wash gold.
	Mien C.	- - - - - -	Coarse wash gold.
	" "	Ngan H.	Nugget gold N. E. at Kinshan.
	Ningyuen F.	Yenyuen H.	Gold 30 li W. at Hokinhoshan, and very coarse gold 150 li N. W.
	Pauning F.	Kwangyuen H.	Coarse wash gold.
	" "	Pa C.	Coarse wash gold.
	" "	Kien C.	Coarse wash gold.
	Chungking F.	Yungtsang H.	Gold washings.
	" "	Hoh C.	Gold washings.
	" "	Fuh C.	Gold washings.
	Yuyang C.	Pangshui H.	Coarse wash gold.
	Chung C.	- - - - - -	Coarse wash gold.
	Kweichau F.	Wan H.	Coarse wash gold 3 li S.
	Suiting F.	Tatsoh H.	Gold.
	Lungngan F.	Pingwu H.	Coarse wash gold.
	Mei C.	- - - - - -	Coarse wash gold.
	Lu C.	- - - - - -	Coarse wash gold in the Tsungkiang R.
	Ya C.	- - - - - -	Coarse wash gold in the Fihkiashui R.
	Mau C.	- - - - - -	Gold.
Chehkiang.	Ningpo F.	- - - - - -	Gold at Kehyüshan.
	Yenchau F.	In all the H.	Wash gold.
	Chuchau F.	Lungtsiuen H.	Light-colored gold.
	- - - - -	Sungyang H.	Light-colored gold.
Fuhkien.	Taiwan F.	Fungshan H.	Gold E. at Kinshan.
	Fuhchau F.	- - - - - -	Coarse gold.
Kiangsi.	Nanchang F.	Fungsin H.	Gold-sand.
	Jauchau F.	Poyang H.	Gold at Hwangkingtseh.
	Fuchau F.	Lingtse H.	Gold 40 li W.
	Kanchau F.	Shuikin H.	Gold.

[1] Peking. [2] Barkoul.

TABLE OF GOLD WASHINGS AND MINES.—*Continued.*

Province.	Department.	District.	Place and circumstances of occurrence.
Kwangtung.	Shauchau F.	Yingte H.	Gold.
	Hwuichau F	Hoyuen H.	Gold at Lantienta.
	Shauking F.	Kaikien H.	Gold at Kintsung.
	" "	Kwangning H.	Gold at Kinkung.
Hunan.	Changsha F.	------	Gold.
	Hăngchau F.	------	Gold.
	Yuenchau F.	------	Gold.
	Changteh F.	------	Gold.
	Chin C.	------	Gold.
	T'sing C.	------	Gold.
	Yochau F.	------	Gold.
Kwangsi.	Liuchau F.	Yung H.	Gold.
	" "	Laiping H.	Gold.
	Sz'ngan F.	Pin C.	Gold.
	" "	Tsienkiang H.	Gold.
	" "	Shangling H.	Gold.
	Pingloh F.	Pingloh H.	Gold.
	" "	Yungngan C.	Gold.
	Wuchau F.	Hwaitsih H.	Wash gold in river at Kinngohshan 70 li W.
	Sinchau F.	Kwei H.	Gold.
	Nanning F.	Hwang C.	Gold mines.
Kweichau.	Tungjin F.	------	Gold-sand washings 100 li W. in the Sungchi R., and 140 li W. in the Tichi R.
	Tsuni F.	Tungtsz H.	Gold.
Yunnan.	Tsuhhiung F.	Yau C.	Coarse gold in the upper Tayauho R.
	" "	Tsuhhiung H.	Gold in the Yenshan.
	Likiang F.	------	Gold washed in many places in the Kinshakiang for a distance of 500 li.
	Yungchang F.	------	Gold mines in the Changpangshan.
	" "	------	Gold washings in the Lantsan R.
	Tungchuen F.	------	Gold washings in the Kinshakiang.
	Yungpeh (Ting)	------	Gold.

Before attempting to sketch the distribution of the known formations of the Chinese empire, I will give the principal reasons for assuming a general simplicity in the geological structure of that country; for believing that the surface of the Eighteen Provinces is made up almost exclusively of the following formations: the Granito-metamorphic,[1] the Devonian limestone, the Triassic, Coal measures, and the younger Tertiary and Post-tertiary deposits.

Wherever the rocks beneath the Devonian limestone were seen, in central and in northern China, these were found to be either metamorphic schists, or granitoid rocks, with the one exception of a thin bed of sandstone, already mentioned as underlying the limestone at the entrance to the Lukan gorge of the Yangtse. At the Meiling pass, on the northern boundary of Kwangtung, the limestone is said to rest on granite.

An exception to this rule exists, perhaps, along the coast range in southeastern China, where the valley of the Canton river is said to expose an extensive formation of "graywacke" resting on granite.

---

[1] By the Granito-metamorphic formation is here meant the stratified and non-stratified rocks of different ages, older than the Devonian limestone.

The Sinian, or N. E. S. W. system of elevation corresponds in many respects to our Appalachian system, and if the analogy holds good throughout, it seems probable that the Sinian revolution terminated soon after the deposition of the Chinese Coal measures, a supposition that is corroborated by the absence, so far as my observation goes, of any younger formations elevated by this revolution.

The apparently total absence, in the line of the Yangtse, of eruptive porphyries, greenstones, trachytes, and basalts, seems to point to a corresponding absence of subsequent disturbance through a large area of the country.

Again, were there fossiliferous strata of the Jurassic or Cretaceous ages, their petrifactions would be found in all parts of the empire, used as curiosities and as medicines, as is the case with the fossil brachiopods and orthoceratites. This is important evidence in China, where art is based on the remarkable, or rather strange, in nature.[1]

In classifying the above tabulated data, I have assumed that the gold washings are indicative of the neighborhood of the granito-metamorphic formation, and have referred this to the adjacent ridges. I have also assumed that the limestone marble, lime, caves, stalactites, and fossil brachiopods, etc., all point to the presence in each locality of the same great bed of Devonian limestone. My own observations in the northern provinces and along the Yangtse, those of Blackiston in Sz'chuen, and the remarks of casual travellers in the south, all point to one, and only one, great limestone formation, which everywhere underlies the coal-bearing rocks, and to which, in all probability, all the indications above given refer.

That the brachiopods belong to this formation is merely an inference, for I never was able to find a fossil of any kind in the limestone. It is, however, an inference based on circumstantial evidence, as when they are frequently cited as occurring in caverns or in the same neighborhood with marble, or stalactites, etc., or in close proximity to coal localities.

With regard to the coal-bearing rocks, I have supposed the coals to belong to the same age throughout the empire, excepting a few which seem, from their names, to be tertiary brown coals. The similar character of the fossils, from the north and from the Yangtse, and the position of given localities with reference to the limestone in many parts of the country, favor the assumption.

Had we good topographical maps of China, the sketch I am about to attempt would be much facilitated; but although the water-courses are laid down on the Jesuit map, with a general approximation to accuracy that is very remarkable, we have very little knowledge of the orography. In the first pages of this paper I pointed out the prevalence of the northeast, southwest direction in the prominent features of Eastern Asia, and went so far as to apply this rule to the establishing

---

[1] Both the Chinese and Japanese have a strong taste for the *bizarre* in nature, as shown by their fondness for dwarfed or deformed trees. Waterworn and cavernous rocks are carried long distances to be used in ornamenting gardens, and quarries are worked for blocks of dendritic limestone to be made into articles of furniture or ornament. All kinds of fossils are esteemed as medicines, and sold as such in all apothecary shops, the brachiopods as Shïyen "stone swallows," and the fossil bones and teeth, from caverns and loam deposits, as "dragon's teeth," "dragon's scales," "dragon's bones," etc.

of several principal anticlinal axes of elevation in China Proper. In this sketch I shall endeavor to give more reasons for the locating of these ridges, which, on the small, general sketch-map, are represented by the limestone and granite streaks.

In describing the structure of the northern part of Chihli and Shansi, a range was often mentioned under the name of the Barrier range. Its trend is here west of S. W., and its prolongation would cross the Hwang Ho in Pauteh (chau), and thence run S. W. through Shensi and Kansuh, coinciding with the watershed between the eastern and western reaches of the great bend of the Hwang Ho. We have already seen that this range has elevated the Devonian limestone in its northeastern part. The Hwang Ho traverses it through an immense gorge, a fact which in China is almost proof of the presence of the limestone. West of this range are the coal localities of the Ninghia (Fu) and Lanchau (Fu).

The next great axis, to the eastward, seems to originate, like the former, in the mountain-knot of the Ourangdaban, near the Tushi gate of the Great Wall, N. W. from Peking. Following a S. W. course it forms the range which we crossed at the Nankau pass, and crossing the Shansi boundary it is known as the sacred Wutaishan. Still further to the S. W. it crosses the Hwang Ho under the name of the Lungmun shan [mountains of the Dragon gate]. In northern Chihli we have seen that this is a granite range flanked with the Devonian limestone; the latter formation is indicated to the S. W. in the lime works west of the Făn river, in the caverns of Taning H. and the lime of Kih C., in the celebrated Lungmun gorge, through which the Hwang Ho passes this range and in the caverns of Fungtsiang F. I have supposed its continuation bordering on the highlands of western Sz'chuen, forming the watershed between the Sz'chuen and Tibetan sources of the Yangtse.

Between these two apparently principal axes there seem to be minor ones, but I have colored the intervening space as Coal measures. In it lie the coal basins of Siuenhwa F. in Chihli; of Tatung F. and Tsingloh H. in Shansi; and of Yulin F. and Pingliang F. in Shensi.

We come now to the central axis of elevation, to which attention was called in the beginning of this paper, and the establishing of which was there based on a study of the map. Where this range crosses the Yangtse, we have seen that it consists of two anticlinal ridges of limestone with an aggregate breadth of 80 miles, and containing between them a coal basin. In its continuation S. W. to the Nanling mountains it seems to occupy a large part of Kweichau. The only data for this portion of the range are, the numerous gold washings at the base of the watershed between Kweichau and Hunan, that I have taken as indications of the granito-metamorphic formation, and the caverns and marble localities of Shihtsien F. and Chinyuen F. In its continuation to the N. E. it is crossed by the river Han, and gives rise to the sources of the Hwai river. It disappears at the edge of the great delta plain to rise again as the watershed of Shantung. In this province the numerous gold localities that stretch through the centre from S. W. to N. E. indicate the presence of the older metamorphic rocks, which, indeed, according to my own observation, form the coast near Chifu. The stalactites of Taingan F. and Kü C. are the only data for coloring in the limestone. The continuation of this range further to the N. E. is found in the limestone islands that stretch from Shantung to the "Regent's Sword," and thence through Liautung, as the Changpeh shan, dividing the waters of the Yaluh and of the Usuri from those of the Liau and the Sungari. In passing close under the precipitous shores of Liautung, I observed that this promontory is made up of parallel N. E. S. W. ridges, and the rocks had all the appearance of limestone.

Between this central axis and that previously described, lies, perhaps, the most important fold of the Coal measures. Beginning in the extreme north, we find coal at several localities along the west coast of Liautung, and along the "Palisade" west of the Liau river. In northern Chihli are the coal basins of Yungping F., of Peking, and of Kwanping F.; in Shansi those of Pingting C., Taiyuen F., Fănchau F., Hoh C., Pingyang F., Tsehchau F., and Kiang C.; in Honan those of Honan F. This main fold, or zone of folds, seems to occupy a large part of the provinces of Sz'chuen and Yunnan. Many minor ridges bring the limestone to the surface in these provinces. In this region almost all the indications of the Coal measures, exclusive of the information given by Capt. Blackiston, refer to the great salt deposits. The following considerations have led me to look upon these deposits as members of the Chinese Coal measures. Some, at least, are in the neighborhood of abundant coal mines.[1] Thick coal seams are sometimes bored through before reaching the salt. They occur at various points along the Yangtse as in Wushan H., Chingking F., and Süchau F., in all which places they must be very near ridges of limestone, but above that formation. In Shunking F. and in Kiating F., they are also near such ridges. If the wells are in rocks younger than the limestone, their depth (500 to 2,600 feet) cannot penetrate to anything older than the limestone. This, and the fact that thick seams of coal are bored through in these wells, and the remark of Blackiston that all the coal rocks he saw in Sz'chuen resembled those of the Kwei coal field, the character of which we know, render it, I think, probable that both the coal and the salt deposits belong to the Chinese Coal measures.

The region in question, though containing many small parallel troughs, seems to be, as a whole, a major trough, if I may use the expression, between two principal anticlinal axes, and, as such, it seems to be traceable through Eastern Asia. To it the S. W. N. E. course of the Yangtse in Sz'chuen owes its direction, and the same may be said of the northern part of the delta plain, the Gulf of Pechele, the valley of the Liau river, and that of the lower Amur, and the depression in which lies the Gulf of Penjinsk.

On the sketch map the two members of the central anticlinal axis, which we have seen to exist where it crosses the Yangtse, are represented as continuing separately in Honan and Kweichau. Whether the course of the Wu river, in the latter province, is sufficient indication of a continuation of the synclinal trough of Kwei toward the S. W. is doubtful, but to the N. E. the coal basins of Ju C. in Honan, and of Yihte H. (Tsingchau F.) in Shantung fall in that line.

East of this central axis is another major trough or basin. In this are some of the coal basins of Hunan, the lake-plain of the Tungting, and the valleys of the

---

[1] Imbert, in Annales de l'Assoc. pour la propag. de la Foi.

rivers Yuen and Tsz, all in Hunan, and in Nganhẁui the valley of the Hwai, and the coal basin of Süchau in Kiangsuh.

This trough is limited on the east by what would seem to be a band of parallel ridges extending from the province of Kwangsi to Kiangsuh. We have seen the Yangtse crossing one of these between Hankau and Kiukiang, while another, broken through by the Poyang lake, shuts in the valley of the Yangtse on the east. The river flows between these two from the Poyang to beyond Nanking.

Numerous indications of the limestone as stalactitic caves, fossil brachiopods, etc., extend in a southwest direction through Kiangsi and Hunan into Kwangsi, while in the same belt are many evidences of the Coal measures.

That the space between these ridges is occupied by coal basins in part of Kiangsi and Nganhwui is certain, and here belong also the coal basins of southeastern Hunan. I have, therefore, represented them as independent throughout. In the easternmost of these, east of the Poyang lake, are the granite hills of Kingteh, which furnish the celebrated kaolin[1] for fine porcelain, while Abel mentions granite and micaceous schists as occurring in the high hills west of the lake in the western ridge.

The data for the next trough to the east are the existence of what seem to be shales and sandstones of the Coal measures on the Kan river from Nanchang F. to the Meiling pass, and the coal fields of Kwangsin F. (Kiangsi), of Küchau F. and Chuchau F. (Chehkiang), of Ningkwo F. (Nganhwui), in every *hien* of which there is coal, and of Huchau F. (Chehkiang).

We come now to the coast axis of elevation marked by the range of mountains that separate Nganhwui and Kiangsi from Chehkiang and Fuhkien.

We know that at the Meiling it is of granite flanked with limestone; the fact that Mr. Fortune found the peaks near the headwaters of the Min river to be granitic, and in the northeast the granitic islands of Chusan, all indicate a granite range, while the table furnishes numerous evidences of the presence, on both sides, of the great limestone formation.

There are even fewer data for understanding the structure of the eastern and southern provinces than for almost any other part of the empire. Scattered indications of limestone and coal, and the courses of some of the rivers have prompted me to insert another axis of elevation, nearer the coast and stretching from Hongkong to Wanchau F. in Chehkiang. Such an axis is apparent in the granite[2] islands that stretch away toward Hainan, and to it this island seems to belong.

The indications of the Coal measures along the coast are the coal fields of Hinghwa F. and Nganki H.[3] (Tsiuenchau F.).

The prolongation of the coast axis of elevation cuts the southern and most mountainous parts of Corea, and coincides nearly with the granite axis of Kamschatka.

I have thus far in this sketch made no mention of any other system of elevation than the N. E. S.W.; but, as we have seen in a former chapter, another system, the

---

[1] This word is said to be derived fron *kao*, high, and *ling*, ridge.

[2] Chin. Repository.

[3] This I take to be the *Anko* mentioned in the Chin. Rep. as producing anthracite.

E. W., exists, and to its disturbing influence are due some of the most important and beneficial features in the structure of the country.

Between the Wei river of Shensi and the Sz'chuen boundary, two ranges, parallel branches of the prolonged Kwenlun, with a general trend from west to east, penetrate far into Central China. Some of the peaks of these chains are said by Klaproth, on Chinese authority, to rise above the snowline. The numerous gold localities in this region point to an extensive development of the older metamorphic rocks, while the presence of stalactitic caves and other indications of limestone seem to show that this formation flanks the ranges in question.

The trends of the upper courses of the rivers Han and Kialung, and the communication said to exist between these streams at Ningkiang C. seem to indicate that the space between these ridges is an elevated table-land, divided by a low watershed that separates the sources of the Han from those of the Kialung. This watershed would be in the line of the limestone range represented as crossing Shansi, Shensi, and Western Sz'chuen.

The disturbances caused by the northernmost of these ridges ceases in Honan, but the southern member seems to continue farther east, apparently crossing Hupeh into Nganhwui.

Of the mountains in Southern China that belong to this system, we know as little as of those just mentioned. They are spoken of as containing snow-capped peaks and high table-lands in Kwangsi and Kweichau, and are supposed by Humboldt[1] to be the continuation of the Himalaya mountains. The hydrography of Yunnan, as shown on the great map of Kanghi, would seem to indicate the existence of a more or less elevated plateau, which, beginning west of the Lantsan river, trends nearly east, entirely across Yunnan, occupying a region in which rise tributaries both of the Yangtse and the Si Ho, and of the rivers that flow to the Gulf of Tonquin. The little that is known of the climate of the city of Yunnan F. (in about 25° N.) tends to confirm the supposition that it is on an elevated table-land.[2] This plateau seems to extend to the western part of the province, where it appears to terminate abruptly toward the plain of the Irawaddi river, for Marco Polo required two days and a half to descend from the city of Yungchang F. to the lowlands of Ava, and speaks of the descent as being very great ("grandissima discesa.")[3]

Toward the east these highlands are represented by Klaproth as forming two diverging ranges of mountains, the northernmost of which is crowned with snowy peaks and glaciers till near the head waters of the Yuen river.[4] There seems to be little doubt that in the meridian of Kweilin F., and to the east of that point, this northern branch forms a comparatively low range, and is nearly lost in the N. E. S. W. system.

---

[1] Asie Centrale. [2] Ritter, Asien, III, 754. [3] Ritter, Asien, III, p. 746.
[4] Ritter, Asien, III, p. 660. Klaproth, Mag. Asiat., II, pp. 139, 156.

CHAPTER VII.[1]

THE SINIAN[2] SYSTEM OF ELEVATION.

I HAVE taken the liberty of giving this name to that extensive N. E. S. W. system of upheaval which is traceable through nearly all Eastern Asia, and to which this portion of the continent owes its most salient features.

We have seen how generally prevalent this trend is in China, whether we consider the hydrography, the courses of the mountains,[3] or the strike of the strata.

In crossing the plateau of Mongolia from the Great Wall to Siberia, I found the same trend predominating in the uplifted strata of old metamorphic rocks, and generally in the ridges that cross the steppes of the Gobi.

A glance at any recent map of Siberia will show that the same rule may be applied to all of the eastern part of this vast region. The Yablonoi, Altan-kingan, and Stanovoi mountains, with all their intermediate, parallel ridges, that together form the valley network of the upper Lena and Amur rivers, are instances of the development of this system on a grand scale. Although exceptions—that may or may not belong to this system—to the general N. E. trend seem to exist in the Great Kingan mountains—the eastern edge of the great plateau—and in the continuation of the Stanovoi in the far northeast, still to the configuration arising from the prevalence of this trend, are due the most marked features of Eastern Asia. The seas of Ochotsk and of Japan, the gulfs of Pechele and of Tonquin, are geoclinal valleys of this system of great geological age, which the disturbances of a long range of time have not been able to obliterate. And a similar valley is, I think, indicated for the land by the line of reference I have drawn through the valleys of the Yangtse and Amur. As throughout China and across Mongolia I was unable to find anything more recent than the Chinese Coal measures affected by this uplift, and as, to the extent of my knowledge, no younger rocks are affected by it in Siberia,[4] it seems proper for the present to refer all the N. E. ridges to one system, and their origin to one revolution.

The, in many places, unconformable strikes and dips of the older metamorphic schists of China show the existence of disturbances that had ceased before the formation of the great bed of limestone.

---

[1] See Map, Pl. 7.

[2] From Sinim, the name applied to China in the earliest mention made of that country.—Isaiah.

[3] That the general trend of their mountains is N. E. was known to the early native writers.

[4] The explorations of M. Tchihatcheff, in the Altai, the eastern part of which belongs to the system in question, failed to discover any rocks more recent than the Permian, affected by this uplift.

The Sinian revolution seems to have begun after the deposition of the limestone, and before that of the Coal measures; at least the difference in character that is visible between the beds that overlie the limestone on the two flanks of the anticlinal ridge in Western Hupeh, and the presence, at the bottom of the Coal measures near Peking, of conglomerates, formed from porphyries that are younger than the limestone, are facts that seem to favor this idea. It is not improbable that these first movements determined the outlines of the principal areas of land and water, and of the future coal basins. The revolution does not seem to have reached its climax till after the Coal measures had been deposited, when the strata were plicated and prepared for metamorphism.

Very striking analogies are apparent between the Sinians and our own Appalachians. Both have the same trend; both are the results of revolutions, which, though they may not have been coextensive in time, were contemporaneous through a long period; and both have folded immense areas of coal-bearing strata. As the elevation of the Appalachians determined the outline of Eastern America, so the Sinian revolution fixed the eastern boundary of the great continent.

We have, in this analogy, one more link in the chain of evidence toward proving the subordination to harmonious laws of the causes that have produced all the varied features in the configuration of our planet.

One of the most remarkable features in the configuration of the northern hemisphere, seems to me to be the number of geoclinal valleys having a nearly N. E. S. W. course, that characterize it. In the extreme east of the great continent we find one, occupied by the sea, between the Japanese Islands and the coast range of Manchuria; between this and the Kingan mountains[1] another, which I have several times alluded to as the principal line of reference in treating of the Sinian features; the Gobi, including the region between the Kingan and the Altai, forms a third. These troughs have all been referred to in the preceding pages, but, if I may be permitted to generalize beyond the closer limits of this paper, I think a much larger one exists in the vast extent of lowlands that stretch unbroken, excepting by the Ural mountains, from the Altai to the Scandinavian peninsula.

---

[1] The eastern edge of the plateau, unlike the southern, is formed by parallel ridges trending between N. E. and N. by E., the valleys between which form succeeding terraces from the plateau to the Sungari river. Prince Krapotkin, who travelled in disguise from the Argun river to Mergen, ascending the Gan river, and descending the Noumin river, gave me the following information: The ascent to the edge of the plateau from the west was hardly perceptible, the descent to the east rapid. In descending he crossed four parallel ranges trending N. N. E., all of which are traversed by the tributaries of the Sungari. The specimens brought back by Prince Krapotkin, chiefly from the ranges, were mostly granite, porphyries, argillaceous and micaceous schists, and gneiss. Coal is abundant along the eastern slope.

According to M. Radde the mean height of the Amur between the Kingan mountains and the Bureja mountains, is 800 feet above the sea; between Mochada and the Kur river, from 400 to 500 feet.—*Radde, in Petermann's Mittheilungen*, 1861, pp. 449—457.

MM. Saurin and Murray, of the English Legation in Peking, informed me that in ascending to the plateau from the region west of Jehol, they followed a valley through a mountainous district, and reached the table-land without seeing any signs of an abrupt wall, such as it presents along its southern edge.

Through this broad tract two minor valleys are indicated, one in the trough that contains the Aralo-Caspian depression and the lakes of the Barabinsky steppe, and the other containing the Kara sea, the White sea, the lakes of Finland and the Baltic.

Beyond the mountains of Norway the great depression occupied by the Sea of Greenland and the North Atlantic, is one of the best defined in this series of valleys.

Finally, in the vast extent of lowlands of British America we have a great geoclinal depression lying between the Appalachians and the Rocky mountains, forming an elevated geoclinal valley between N. E. and N. W. systems of elevation; just as in the North Pacific Ocean we have a depressed valley of the same kind between N. W. and N. E. systems—the Rocky mountains and the Sinians.

Both Prof. Guyot and Prof. Dana have demonstrated the fact that the principal continental outlines are referable to N. E. and N. W. systems of trends.

CHAPTER VIII.[1]

GEOLOGICAL SKETCH OF THE ROUTE FROM THE GREAT WALL TO THE SIBERIAN FRONTIER.

THE route, here described, after following for about 100 miles that along which the measurements of MM. Fuss and v. Bunge were made, leaves this and remains about 60 miles to the west of it for most of the distance, joining it again in about latitude 47° N.

The journey was made in the months of November and December, the thermometer ranging from +15° to —28° F., with an almost incessant, strong, north-west wind. This, and the fact that we travelled seventeen hours a day, will, I think, be a sufficient excuse for the meagreness of the information. Nothing but the absence of all geological observations over this immense region, prompts the insertion of the following scanty notes.

Nov. 21, 1864. Leaving Kalgan we ascended to the plateau by the Tutinza road.[2] For the first two or three days the intensely cold winds made it impossible to take notes. The great volcanic formation, which we have seen forming the southern edge of the table-land for a long distance to the westward, extends from thirty to fifty miles in this direction, as the only rock in place, and the conformation of the surface is similar to that with which we have become acquainted in describing the journey to the west, only the valleys are generally broader and more shallow.

During the next fifty miles our route crossed several low ridges, chiefly granitic, the intervening plains being covered with the detritus of quartz and metamorphic sandstone. This is succeeded by a rolling country with hills of red granite, diorite, and greenstone porphyry, which continues to beyond the low granite ridge of Mt. Ugundui.[3] The fragments on the surface of the plains were mostly of granite and quartzitic sandstone, together with scattered pieces of lava and pebbles of chalcedony, agate, etc.

Nov. 26. After passing Mt. Ugundui the character of the country underwent a marked change. Our road lay, from the last-named mountain to the Mingan hills, through a depression. In the distance the flat outline of the plateau was seen on all sides, the intervening country being cut up into isolated knobs and ridges by numerous water-courses and lake beds. The structure of the knobs shows them to

---

[1] See Section on Pl. 7.

[2] This portion of the road, as far as the summit of the plateau, was described in a previous chapter.

[3] Many of the names of places, etc , used in this sketch are given on Klaproth's large map of Central Asia.

be the remnants of a deposit the horizontal beds of which were continuous over the area in question. I examined one of these hillocks, about 50 feet high, near lake Bilika Noor, and found it made up of the following beds, from younger to older:—

Compact, yellowish-gray limestone, with a tendency to oolitic structure.

Thin bed of dark clay, or earth, with concretions of manganese.

Bed of finely crystalline, white, saccharoid gypsum.

Gypsum in massive, transparent crystals associated with more or less red clay.

The stratification is horizontal throughout, and the same structure seemed to be continuous as far as the Mingan hills. What the character of the plateau is I could not determine; as seen in the distance it limits the depression with a cliff and long talus.

An alluvial deposit of red loam is present in many of the valleys, and is, perhaps, nearly contemporaneous with the erosion of the water-courses.

Nov. 27. In the morning we found ourselves in the Mingan hills, apparently an isolated protuberance rising only a few hundred feet above the plateau. The rocks of these hills, where first observed near the southern edge, were chiefly quartzite, compact sandstone, and a talco-argillaceous schist, in highly inclined strata trending N. W. and dipping to N. E. Several miles further to the northwest we came to ridges of limestone, in beds also highly inclined, with a strike W. N. W. and dip to S. S. W. This rock resembles the limestone of the hills west of Peking. It is traversed by dykes of greenstone. In the Mingan hills I found a few rolled fragments of basaltic lava similar to that of the southern edge of the plateau.

To the west of these hills lies the broad deep valley of Olannoor, which seems to connect the depression south of these hills with the great plain of Tamchintala, to which we now descend. As we enter upon this steppe we see before us nothing but an unbroken sandy and gravelly plain with a little scattered grass. A considerable percentage of the pebbles on the surface consists of agate, cornelian, and chalcedony.

Nov. 28. The morning found us still travelling on the Tamchintala, but we soon descended into a large valley-like depression. The plateau is here cut into to the depth of perhaps 150 feet, the vertical wall giving an insight into its local structure. The whole exposed thickness consists of horizontal strata of white calcareous sandstone with thin beds of arenaceous limestone interstratified. At the bottom of the section a bed of red arenaceous clay crops out. The sandstone varies in grain from a fine grit to a fine conglomerate, the ingredients of both being apparently identical with those of the gravel on the surface, between which and the underlying rock there is no line of demarcation. If the pebbles of agate, cornelian and chalcedony are derived from the amygdaloidal lava, so common farther south, their occurrence in this deposit throws light on the relative ages of the two formations.

After crossing this valley depression, which is several miles broad, we ascended to the plain at about the same level, apparently, as on the other side.

Nov. 29. During the previous night we left the plain and entered a rough and very undulating country. Here a belt of older rocks, about seventy miles broad, seems to rise a little above the general level of the plateau. Its position is marked on most maps by the boundary line between inner and outer Mongolia.

As we entered these hills during the night I could not see the structure of their southern edge, but where first observed, several miles from that point, the outcropping rock is a compact hard sandstone, in nearly vertical strata trending about E. W. Beyond this the next rock observed was granite in red and white varieties, traversed by numerous dykes of brown porphyry with bright red crystals of feldspar.

The surface of this granite region forms numerous depressions, the bottoms of which seem to be occupied, in the wet season, by ponds without outlets. In the gravel of one of these depressions I found a slightly rounded fragment of silicified wood.[1]

Nov. 30. The morning of this day found us still in the hilly region. The rocks along the road were clay schist. We came, early in the morning, to a narrow gravelly plain, which, descending between two granite cliffs, opened out on to the broad plain of the valley of Ulannoor.

The hills on either side of the narrow plain just mentioned, which are of coarse granite traversed by a similar rock of finer grain, are bare, without either soil or vegetation, excepting two or three dwarf trees growing from crevices in the rock. These trees were the only ones seen on the plateau between Kalgan and the hills of Urga.

Entering the valley of Ulannoor near Gashun we found ourselves in a country of high terraces, these consisting, where seen, mostly of beds of clay. This clay would seem to be the equivalent of the calcareous sandstone, and is covered, in the narrow valley mentioned above, by a deposit of loam.

Crossing the valley of Ulannoor, we entered a valley in the hills of Ulandzabuk-daban. Here the ground was covered with angular fragments of clay-slate, and gneiss.

Rolled fragments of porous lava were also found on the surface.

Dec. 1. This day our road lay through the hills of Senji, which consist of alternating vertical strata of micaceous, argillaceous, and talcose schists, and compact limestone in blue, black, and white varieties, all having a very regular trend to about N. E. These strata are traversed in all directions by dykes of greenstone. Large lenticular masses of quartz were also observed, and some broad veins of the same material, apparently interstratified, and discolored with the oxides of iron and manganese.

The frequent repetition of the more easily recognizable rocks would seem to show a highly folded condition of the strata.

The limestone having better resisted the action of disintegration, forms ridges from 100 to 150 feet high above the bottoms of the troughs formed by the removal of the intervening softer rocks. Thus the general appearance of the surface is that of parallel valleys and ridges. But here too we find the same tendency to form depressions without outlets, that we have already seen in the granite region (Nov. 29th), and

---

[1] Silicified wood was shown to me in Peking under the name of Hanhaishï. Hanhai is the Chinese name for the Gobi desert.

which is mentioned in a previous chapter as occurring along the southern edge of the plateau, in the erosion of the lava region. In all these instances the depressions are entirely in the solid rock, and vary in size from a few yards to several thousand feet across. They have the appearance of being produced by erosion and not by sinking. In the instance before us this conformation is often assisted by cross dykes of greenstone. But the occurrence generally would seem to arise from inequalities in the texture of the rock. Whatever the cause of these depressions may be, their manner of formation is probably closely connected with the origin of a large class of desert lake beds.

For many miles the surface of the rock was entirely bare of soil, excepting in the bottoms of the depressions just mentioned, where ponds are probably formed in wet years.

From this hilly region we came gradually into another of those broad plains, which form, in the aggregate, the true plateau. These plains, the steppes of the Russians, and *tala* of the Mongols, are like those of our own deserts in the Rocky mountains. They are great valleys, often from twenty to sixty miles broad, filled with marine deposits that have retained their horizontal position and remained often intact from erosion. Their surface is not, strictly speaking, horizontal, but slopes from both sides to the centre.

The deposit forming the substructure of this plain, seems to be the same sandstone and conglomerate that we have seen on the Tamchintala, judging from some blocks of these rocks seen near a Mongol dwelling.

Crossing this plain we came, near its northern edge, to a line of basaltic cones from 100 to 150 feet high, isolated from the low flat hills to the north, and apparently resting on clay slate. They seemed thus to belong to a bed or stream rather than to a dyke. Whether the flat hills near by are a continuation of the same volcanic rock I could not determine.

The rock is a brownish-black, minutely crystalline basalt. On the surface of the plain, near these hills, I found large numbers of fragments of black and red cellular lava, and abundant angular pieces of chalcedony, and red and green jasper, etc.

Dec. 2. During this day we crossed two broad valley depressions, the same calcareous sandstone and conglomerate already mentioned, forming apparently the substructure both of the long valley slopes and of the higher land intervening between these. A few fragments of blue limestone and white quartz, derived probably from the formation we crossed yesterday, were found in the surface gravel; but a large percentage of this gravel consisted of chalcedony, cornelian, and agate.

From the highest ground the flat outline of the plateau was visible in every direction, excepting to the south, where we could see the hills of the past two or three days rising to the height of perhaps 1000 feet above the neighboring plateau.

Dec. 3. We travelled the past night and this day on the continuation of the steppes of the last two days. During the afternoon the plain descended gradually to the north till it ceased abruptly against a granite ridge from 50 to 100 feet high. Beyond this ridge, for a few miles, the country though somewhat lower than the plain of the morning, is bare of the steppe deposit, and presents a rough, granite surface.

Dec. 4. Detained one day by a *bouran* or snow-storm of great violence.

Dec. 5. Travelled over a rolling country chiefly of granite and mica schist. Associated with the latter rock is a white dolomitic limestone in apparently interstratified beds, impregnated with specks and flakes of graphite. The general trend of these rocks appeared to be to the N. W.

The granite had, in places, more the appearance of a metamorphosed conglomerate breccia than of a true granite.

In the afternoon we encamped among outcrops of trachytic porphyry identical in character with that of Kalgan. I found here all the kinds seen at Kalgan, including a striped variety, and specimens with primary quartz. This porphyry contains veins and concretions of chalcedony and cornelian.

Dec. 6. Our road lay all day over a rolling country, granitic and syenitic rocks prevailing, till in the evening we reached the foot of a picturesque granite peak, the Bogdo oola,[1] rising several hundred feet above the surrounding country. To the west of this we saw a large valley with water or, rather, ice.

An accident detained us here till the next afternoon.

Dec. 7. Started in the afternoon, and after passing the Lamasery of Churinchelu, and travelling a few miles along the foot of the Bogdo oola, encamped for the night.

Dec. 8. Travelled about 20 miles over a rough country. As the ground was covered with snow, I saw but little of its character, the outcrops seen being all granitic.

Dec. 9. This day we were again on the undulating country of the plateau and the great steppe deposit. Near our camping place were many fragments of volcanic scoriæ and of chalcedony.

Dec. 10. Our road was still on the steppe of yesterday, the surface rising rapidly toward the north. The rolled detritus on the surface was mostly derived from mica-schist, and clay slates, and in a ravine I observed the former rock in place. Near this we entered the hills that limit the steppe, and found them to be of basalt, at least as far as the camping place.

Dec. 11. This day found us in the range of hills that, trending S. W. from the Kentei mountains, forms the watershed between the steppes of the Gobi and the valleys of the Tula and Orkhon rivers, whose waters flow to the Arctic Ocean.

The country is here made up of rounded, grassy hills, of about the same height, with valleys remarkable for the regularity of their long, unbroken, cross curves. The hills are of a black, metamorphosed clay schist, and a compact, greenish rock, chiefly feldspar and quartz, apparently a metamorphic greenstone. The strike of the clay rock, where observed, was N. S., and the dip vertical.

The valley bottoms, and the lower slopes of the hills, are covered with a rich, black earth, the deposit showing no signs of erosion. Our camp this night was in the Horteryndaban.

Dec. 12. During at least the greater part of the past night we were descending, and daylight found us in a valley much like that which leads from Kalgan to the plateau, viz., a narrow, gravelly descending plain, inclosed between hills several

[1] Bogdo, sacred, and oola, mountain.

hundred feet high, and remarkable for their pyramidal forms. The fragments of rock, both angular and rolled, that cover the valley, were found to be of green clay schist, the same metamorphic greenstone seen yesterday, and a greenish sandstone.

In the forenoon we reached Urga, also called Kuren, the residence of a living Buddha.

Dec. 14. Left Urga for Kiachta, which place we reached on the 21st December. The country between these places was covered with snow, concealing its geological character. Our road lay through the hills to the eastward from the Orkhon river, crossing its tributaries, the Kara Gol and the Iro Gol.

Through the first two-thirds of the distance the few outcrops seen were of rocks similar to those seen near Urga; at Iro Gol I found chloritic granite.

A great steppe deposit, apparently of loose argillaceous sand, fills the valleys, and, extending over the lower parts of the crests of the ridges, leaves the higher peaks isolated like the islands of an archipelago. This is part of a very extensive deposit which, from its position here, must be continuous through all the lower course of the Orkhon. It would seem to be the same deposit that forms the broad steppe south of Kiachta, and is visible, I think, in the terraces of the Selenga as far as Lake Baikal, and in the tables on either side of the Angara at Irkutsk.

The barometrical measurements of the Russian Academicians, MM. Fuss and v. Bunge, have shown that that part of the continent which they crossed, between the Great Wall of China and the Siberian frontier, south of Lake Baikal, is an elevated plateau, bounded on the N. W. and S. E. by mountain ranges from 5000 to 10,000 feet high, from the sides of which the table-land falls gently toward a broad level region in the centre, the mean height of which is not more than 2400 feet.

The skeleton of the plateau is thus a great geoclinal valley, trending nearly N. E., the basis of which, so far as observed, is formed by granitic rocks, and metamorphic strata, probably of Paleozoic origin, and the inequalities of which have been nearly filled up with more recent formations. Of these latter we can, at present, recognize only three, viz:—

1. The great development of lava along the southern edge.
2. The steppe deposit including the Gobi sandstone.
3. The deposits of loam, mentioned in the preceding pages as covering in places the steppe deposit.

The lava formation is apparently the oldest of the three. We have seen, in a former chapter, how a part, at least, of the southeastern edge of the table-land owes its level surface solely to the great thickness of the volcanic rocks, which have thus been able to fill up the hollows between the ridges of granite and metamorphic rocks. The profile, constructed from the measurements[1] of MM. Fuss and v. Bunge, seems to indicate the existence of a terrace from 3000 to 4000 feet high and about 150 miles broad, that forms the S. E. border of the plateau. It is not improbable that this terrace is due, in great part, to extensive lava flows.

The volcanic rocks of Lake Baikal and of the region to the east, the occurrence

---

[1] Ritter.

of products of this class in place and as scattered fragments at many points on the route across the plateau, and finally the information derived from Chinese authorities concerning the existence within historical times of active volcanoes, among the mountains of Manchuria to the east, and in the Tienshan of the west, all point to a development of volcanic activity, which was formerly coextensive with the area of the present table-land. The remains of this action still make themselves felt in the violent earthquakes that from time to time shake the districts of northern Chihli and the shores of Lake Baikal.

The greater flows of lavas seem to have been predetermined by the fissures of dislocation, formed along the borders of the area that was subsequently to be elevated. Such a fissure we have seen marked by a great fault south of the Lakes Kirnoor and Téhai.

In the present state of our knowledge of this vast region, it is, I believe, impossible to say whether, at the time of the eruption of these rocks, the present depression of the Gobi was or was not under water. That a portion of the southern edge of the plateau was not submerged appears from the fact that where the bottom of the lava formation was visible it was found to rest immediately on the old granitic and metamorphic rocks. This, however, does not preclude the possibility of the existence of undisturbed deposits under the steppe sandstones of the Gobi.

The sea in which the great steppe deposit was precipitated was studded with islands now represented by the ridges and peaks that rise above the plains. The surface of the plains rises everywhere toward these former islands, partly because the deposit in its formation adapted itself partially to the original surfaces of the valleys it fills, and partly from its thickness being increased by the tributary detritus of the islands. The effect of such a combination of circumstances upon the form of the surface, has been discussed in treating of the lake deposits of Northern China. It seems not improbable that the same causes may have operated here as there, in forming many of those lake valleys, the beds of which rest upon the steppe deposit.

The age of this extensive deposit is a question of much interest. If it is contemporaneous with the steppes and terraces of the valley system of the Orkhon and Angara, it seems probable that the sea which left this deposit over nearly all of what is now the plateau, was also contemporaneous, within certain limits, with that great body of water which, extending from the polar ocean to the Caspian, occupied all Western Siberia.

The fact, to which Baron v. Humboldt[1] has called attention, that seals, identical in species, inhabit the fresh waters of the lakes Baikal and Oron (lat. 55° N., long. 119° E.) and the Caspian Sea, seems to refer to that period. The Oron lake is a tributary of the Vitim, and through this of the Lena, in which no seals occur. This circumstance points very clearly to a former water communication between these far separated localities, and the time at which the seals of the Oron became isolated from those of the Baikal and the Caspian falls, perhaps, in the same period with the emergence of the great plains of Northern and Western Siberia, the deposits of

[1] Humboldt, Kosmos, IV, p. 456. Stuttgart und Tübingen, 1858. Pallas. Zoographia Rosso-Asiatica, 1818, p. 115.

which are characterized by abundant remains of the mammoth as well as of *Bos urus* and *Rhinoceros tichorhinus.*

We have seen that although the effects of erosion are generally not very extensive in the steppe deposit, they exist in some places on a large scale. The deeply cut valley in the Tamchintala is an instance, and one that seemed to me could have been caused only by fluviatile action. The erosion in the neighborhood of Bilika Noor, and the presence in the eroded valleys of loam strongly resembling that deposited by great rivers is another instance. This loam was not often seen, indeed it is mentioned in my notes only as occurring in the Mingan hills, at Bilika Noor, and over the steppe deposit near Goshun.

The closing event in the history of the great sea that in comparatively recent times covered so large a part of Asia, extending from the pole to the Caspian and Black sea, and from the Ural mountains to near the Great Wall of China, was the disappearance of its waters from the long trough that reaches from the shores of the Arctic sea, through the Barabinsky steppe to the Aralo-Caspian depression.

It appears to me that the ancient physical geography of this vast region, and the effects of its elevation, present one of the most interesting and important fields of exploration. Whether we consider the meteorological changes that must have been brought about by the upheaval of so large an area, or the influence of this great water communication and its currents on the distribution of existing genera, the geological phenomena that have affected this broad belt of the great continent have, beyond doubt, had an important influence on the recent history of our planet.

In the following table I have recapitulated the few leading events in the geological history of China and Mongolia which seem to be recognizable.

	Disturbances.
A. Deposition and metamorphism of the older metamorphic strata of China. B. Deposition of the metamorpic strata of Mongolia.	Uplifts apparently of various ages and directions, of which the surface effects are mostly obliterated.
C. Deposition of the great Devonian limestone formation.	
D. Eruption of the older porphyries of the Sishan, west of Peking. E. Deposition of the Chinese Coal measures. F. Eruption of the younger porphyries of the Sishan.	Sinian revolution forming the N. E. S. W. system of uplifts. Emergence of all China Proper.
	Submergence of Mongolia.
W. Eruption of the trachytic porphyries of Kalgan and the Gobi Desert. X. Eruption of the volcanic rocks of S. Mongolia and the Baikal region. Y. Deposition of the steppe deposits of the Gobi Desert.	
	Commencement of the emergence of the plateau. Formation of the great dislocation along the southern edge of the plateau. Supposed change in the course of the Hwang Ho, and formation of the chain of northern lakes.
Z. Deposition of the lake loam of the northern lakes. Beginning of the delta of the Hwang Ho.	
	Deepening of the channel of the Hwang Ho between Shansi and Shensi, and of the gorge of the Yang Ho, and consequent drainage of the northern lakes.

CHAPTER IX.[1]

GEOLOGICAL ITINERARIES OF JOURNEYS IN THE ISLAND OF YESSO, IN NORTHERN JAPAN.

THE following notes were taken during journeys made in the service of the Japanese Government, in the summer and autumn of 1862. As the very small population of this northern island is composed almost entirely of fishermen, it is confined to small villages scattered along the sea-shore. The only roads are those connecting these hamlets, with the exception of rare bridle-paths penetrating the interior. The mountains west and north of Volcano bay are covered with dense forests and a denser undergrowth of a kind of bamboo, so close-set that the country is impenetrable, excepting by wading in the beds of torrents.

Thus the geologist is obliged to content himself chiefly with the sections exposed on the sea-shore.

Hakodade, the seat of the Viceroyalty of Yesso and Krafto,[2] is at the foot of a peak about 1,150 feet high, connected with the main island by a low, sandy neck.

The rock that forms this island-like promontory is apparently a pluto-neptunian product resulting from the metamorphism of trachytic tufas and conglomerate-breccias. Where I examined it, it consisted of a fine-grained felspathic base, containing—

1st. Felspar in oblong crystals, from very small to one-third of an inch in length. These were white, highly fractured, and frequently showed triclinic cleavage.

2d. Quartz in pellucid grains, very irregularly distributed, in places absent, in others equalling the felspar.

3d. Hornblende in small prisms.

4th. Magnetic iron in grains.

The rock in this locality has somewhat the appearance of having been broken up and partially refused, but more generally it shows signs of stratification, and I have referred it to the extensive marine deposit formed out of the debris of volcanic rocks.[3]

On the northern slope of the peak is a terrace of recent gravels raised 100 feet or more above the bay.

Between the hills of the main island and the sea there lies a plain the surface of which slopes gently toward the water, where it terminates in places in high bluffs,

---

[1] See Map, Pl. 8.

[2] Sagalin of the Russians.

[3] This is probably the rock described in Com. Perry's Japan Expedition, as granite with crystals of turmaline.

in others in low terrace steps. Near Kameta this terrace is covered with a few feet of clayey sand, underneath which is a bed of whitish clay used for fine tiles; more generally these terraces are a bluish, sandy clay, rich in recent shells, and fringing the less precipitous shores of most of the Japanese islands.

First Excursion. May 24th, 1862.—Leaving Hakodade we crossed to the main island by the low neck of land. This is formed by a bar of stiff clay, perhaps of the same age as the terrace deposit, which lies a few feet above high-water, and is covered with drift sand. Along the eastern edge of the neck, the sand has been raised by the winds into hills, sixty to eighty feet high, the shapes of which change with every storm, excepting where protected by a sufficient growth of wild rose-bushes. Behind these hills the ground is swampy, the water finding a very slow drainage through the sand.

Fig. 9

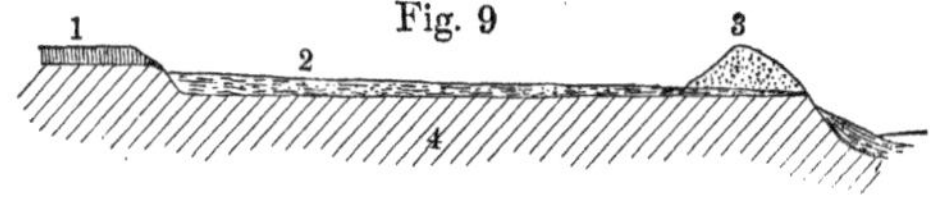

1. Loam. 2. Marsh. 3. Drift sand. 4. Stiff clay.

Following the beach of the northern shore of the bay for several miles, we turned off at a small village, and, ascending a creek, entered the fertile valley of Ono, a broad marshy plain on which are some of the principal farms of the island. An inferior rice and silk are said to be among the chief products.

May 25th.—Branching off from the main road, a few miles beyond the village of Ono, and following a mountain brook, we reached the lead mines of Ichinowatari.

These mines lie at the entrance to a small valley, on the sides of which the out-cropping rocks, containing the veins, are black and gray argillites, slightly calcareous, and highly metamorphosed, in alternating beds; the gray rock being apparently the younger. These are associated with greenstone, whether eruptive or metamorphic was not ascertained, which occupies most of the valley to its head. On the summit of the ridge the greenstone was found by Mr. Blake to be succeeded by a shale, from which he took a calamite, and this again by the black rock already mentioned.

The veins occur in all of the above rocks; the predominating veinstone being of magnesite bearing, in nodules, threads, and impregnations, black and yellow zinc-blende, iron pyrites, galena, and, in places, copper pyrites. The wall rocks are highly impregnated with small cubes of iron pyrites.

In Japan, as in China, the want of pumping machinery prevents working to any considerable depth below the adit level. The galleries in this mine were tolerably well timbered, but low and narrow. From ignorance of the use of powder in blasting, their means of attacking the rock were—till the application of powder in mining was introduced by us—confined to the use of pointed instruments, a miner's pick with one point, similar to our own, a hammer and gad with handle, like the German *Eisen*, completing the outfit. The ore is roughly assorted by hand, and then passed under dry stamps. I was not a little surprised to find, in the mountains of Japan, stamps constructed on the same principle as our own, though the workmanship and efficiency are far inferior.

An overshot water-wheel turns a slender shaft, armed with long cams, by which the stamps are raised. These last are ten in number, of wood, about nine feet long and four inches square, and bear inserted in their lower ends, iron heads from one and a half to two inches square. Each stamp acts in a separate stone mortar, set into the ground, and powders thirty kan,[1] or two hundred and fifty pounds of ore per day of twelve hours. After being stamped the ore is sifted and sent to the wash-house, where it is concentrated to a very pure schlich by hand washing in wooden pans. This work is done mostly by women.

The furnace in which the ore is smelted is a cavity in the ground, lined with charcoal powder kneaded with puddled clay, forming a hemispherical crucible (*b*) about 14 inches broad and 10 inches deep, with an underdrainage. In front is an earthen shield (*c*) to reflect the force of the blast, which enters through a clay nozzle (*d*) from the box-bellows (*e*). The greater part of the smoke, etc., passes off through a large chimney (*a*).

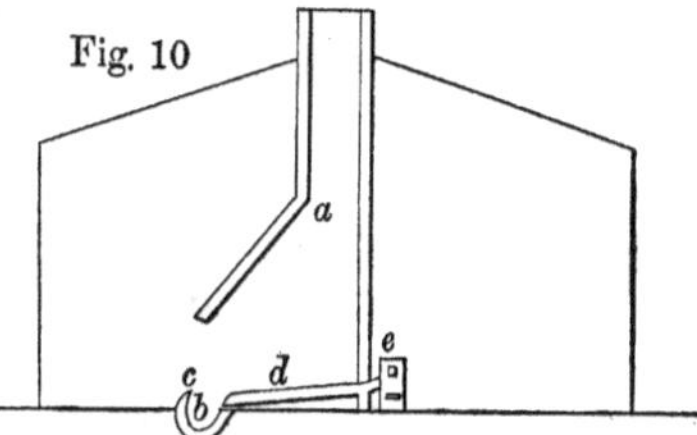

The crucible is lined with charcoal, and when fully dried about 80 lbs. of ore is added and covered with charcoal. When half melted 30 per cent. of pig-iron in lumps of about an inch cube is added. As soon as about one-half of the galena is freed from its sulphur, the whole is stirred. After about two hours the coals are withdrawn, the blast stopped, and water is thrown on the bath to cool the first layer of matte. This is repeated six or seven times till the surface of the lead is free, when it is cast in bars, the matte being thrown away.

We have in this operation the simplest form of the precipitation process, the *Niederschlag Arbeit* of the Germans.

The greatest production at these mines was in 1860, when, during three months, it averaged about 600 lbs. daily; at the time of my visit it was about 80 lbs.

The running daily expenses of production for this small result of 80 lbs. were nearly as follows :[2]—

30 miners, averaging	6 cents	$1 80
30 coolies, at	8 "	2 40
7 overseers, at	5 "	35
1 carpenter		8
26 ore dressers, averaging	3 cents	78
2 stamp tenders, at	4 "	8
1 smelter		8
2 smelter's assistants, at	4 cents	3
200 lbs. of charcoal		17
30 lbs. of inferior pig-iron,		16
		$5 98

[1] 1 Kan is equal to about 8 lbs.

[2] Assuming the ichibu to be worth $0 33.

The miners working in ore are paid according to the weight and quality of the ore extracted, receiving one cent for every 10 kans, or 80 lbs. of best rough ore, and one-half a cent for the same quantity of inferior.

When not working in ore they are paid by the running foot on the gallery and the hardness of the rock, receiving per running shak,[1] or foot, 60 cents for the hardest rock, and 14 for the softest, the average at these mines being 30 cents. One man can advance a gallery one foot, in the hardest rock of these mines, in five days.

The timbering of the levels costs 10 cents per running foot, the wood growing in the vicinity.

May 28th. Leaving the mines, we returned to the main road, and crossed the watershed of the peninsula. The rock is concealed, but judging from numerous fragments on the surface the older rocks of the ridge are covered with volcanic conglomerate.

About twelve miles to the N. N. E. we saw the half ruined cone of the volcano Komangadake, also called the Sawaradake. In the valley lying between us and the peak, lay a picturesque lake surrounded by forests and meadows, and its banks overhung with a rich vegetation. Beyond lay the beautiful Volcano bay. Descending from the ridge we passed the lake, and stopped for the night at the small village of Skunope.

May 29th. Leaving Skunope we started to ascend the volcano. As our way lay through the forest, coolies were sent ahead to clear a path in the underbrush. For several miles we were in a dense wood much like a New England forest; the prevailing trees being grand specimens of magnolia, beech, birch, maple, and oak, with immense vines of grape, ivy, etc., clinging to their trunks and hanging from the boughs.

We came out of the forest upon the gentle foot-slope of the mountain, here covered with a deposit of pumice that extended from where we stood to the summit, in the shape of a stream several hundred yards broad. Leaving the horses, and keeping on the pumice, we soon reached the steeper ascent. The sides of the volcano have been covered with a growth of large trees, where now only dead, white trunks are left, some standing, but the greater number fallen. Many of these lay in our path, while some, standing in their original positions, were surrounded by the subaerial deposit of pumice which reached several feet above the roots.

We reached the edge of the crater at a point below the highest peak.

I was told that the Sawaradake was formerly a single cone, but that seven or eight years before our visit this fell in, the occurrence being accompanied or preceded by a severe earthquake, and an eruption of hot water and pumice, the sand of which was carried by the winds as far as the Kurile Islands.

The crater is now several hundred feet deep, with steep walls, and entirely open toward the sea on the east. The bottom is formed by a convex mass of pumice which extends with an unbroken slope through the opening to the sea-shore.

Great cracks traverse this plain in every direction, distinguishable, from our position on the summit, by their raised, yellow edges, forming long ridges, as though gigantic moles had undermined the plain, and by rows of steam jets

[1] The shak is about one-fifteenth of an inch shorter than our foot.

The view in the distance is grand. On our left the shore of the beautiful Volcano bay forms a long, sweeping curve, parallel to which the mountains in the background, covered with dense forests, appear in all the shades of green, blue, and purple, as they stretch away on the far horizon. Far over the bay, rising as it were from the sea, are several beautiful cones, long quiet, covered to the summits with vegetation, while nearer, though seemingly among them, is the semi-active Usu, a ruined cone whose yellow, sulphur-coated cliffs glisten even at this distance.

We descended into the crater by a talus of pumice, and crossing to the north side came to the edge of a secondary crater, or pit, in the plain. This was about 600 feet in diameter, with precipitous sides on which the stratification of the mass of pumice that fills the bottom of the great crater is distinctly visible.

From the bottom and sides of this pit columns of steam were rising, incrusting the walls with crystals of sulphur and salts. This inner crater must have been formed after the falling in of the cone, and was, perhaps, the point of exit of the ashes that fell after the breaking in of the peak.

On examining the long fissures that traverse the plain, their sides were found incrusted with delicate crystals of sulphur and sulphate salts, while the pumice walls were half turned to a bright red clay, impregnated with these crystals. Putting my thermometer, which was graduated only to 80° C., into the steam, the mercury instantly ran up to that point.

The recent covering of pumice conceals, in most places, the true structure of the mountain, as it forms a deep mantle over every slope not too steep to retain it. This product is grayish-white, very irregular in its porous structure, and contains numerous crystals of felspar and grains of a translucent, greenish glass. It is undergoing rapid disintegration. Bombs of black scoria were found containing crystals of white felspar, and showing transition, in streaks, into pumice characterized by the same contents as that just described.

Blocks of a grayish trachytic lava, abounding in crystals of triclinic felspar and grains of the greenish glass, mentioned above, occur in the crater, and seem to be the rock of which the pumice and bombs are a variety.

The western side of the crater wall is the highest, and owes its better preservation to a broad dyke of rock consisting mainly of a dark paste with greenish-white crystals of triclinic felspar, hornblende, and magnetic iron. The dyke has a tabular structure, the plates being upright in the middle and horizontal on the sides, forming there a right angle with the cooling surface, as is the case with columnar structure. The rock traversed by this dyke was found very much disintegrated.

Without visiting the top of the northern wall we could clearly distinguish the original outer mantle of the volcano, in the exposed edges of different colored strata, while just under the top of the western wall a stratified remnant of what was probably the old cone remained. The greater part of at least the western and northern walls appear to be of trachytic rock.

The general appearance of this mountain produced upon me the impression that it had, before this, been a ruined cone, but was rebuilt by an eruption of pumice to be again broken down and given over to the levelling solfatara-action.

Descending by the same route we returned to Skunope.

May 31st. Leaving Skunope in the morning, we travelled northward, first through a thickly wooded, swampy district, with corduroy road, then over a soil of volcanic ashes, till we finally reached the sea-shore, when turning eastward, we skirted the northern foot of the volcano, and crossing the outlet of the lake reached the fishing village of Shkabe.

The northern slope of the mountain was formerly covered with timber reaching high up its side, and now represented by a forest of dead trunks extending over thousands of acres. The trees were probably killed by the shower of pumice which covered the surface to the depth of from six inches to two feet. On a large proportion of the trees the bark is intact, and they show no signs of the action of fire. A fresh undergrowth was springing up, at the time of our visit, and of this the climbing plants seem to have been the first to start into life.

In the side of a gulley in the bluff, I observed the following series from younger to older:—

1. Layer of pumice, two feet thick.
2. Vegetable mould with roots of grass six inches.
3. Layer of pumice, three to five feet.
4. Thin layers of pumice and sand, apparently an ancient beach.
5. Volcanic conglomerate-breccia.

This section is repeated in all the cuttings observed at the foot of the volcano.

At Shkabe there are several hot springs used for bathing. One of these, rising on the beach and bubbling strongly, has a temperature of 75° C.; and in another rising in a cold stream, but protected by wooden tubbing, I found 70°. The water of these springs has a slight odor of sulphuretted hydrogen.

June 1st. Soon after leaving Shkabe we passed an outcrop of quartziferous porphyry, showing columnar structure, and remarkable for its richness in double pyramid crystals of pellucid quartz associated with white felspar in a compact gray paste. The volcanic conglomerate-breccia was the prevailing rock, but in places the bluff was formed of an apparently younger deposit of sandy clay. The beach was in many places covered with a layer of magnetic iron sand, from the disintegrated volcanic rocks, well concentrated by the action of the surf.

From Shkabe eastward many fragments of vein quartz were seen on the beach.

At the mouth of the Kakumi creek we left the sea-shore, and following the wild valley rode a few miles inland to the mines of Kakumi. Here the hills are formed of greenish and gray argillaceous rocks in places brecciated, in others metamorphosed to an euritic rock. These are traversed by dykes of a peculiar white porphyry.

This porphyry has a compact paste, generally very white, sometimes gray or greenish, yielding fire with difficulty with the steel. In this are scattered grains, and especially double pyramid crystals, of quartz, which form from a few per cent. to one-third the volume. In rare instances it contains crystals of a white triclinic felspar. Mica and hornblende are never present and rarely chlorite. It contains almost always small cubes of iron pyrites.

In weathering it changes to a white kaolin-like substance often discolored by the oxidation of the pyrites.

It occurs in dykes, and often shows columnar structure.

Porphyry of a similar character occurs at several points on the island.

Ascending the creek, greenstone was found to succeed to the argillaceous rock, and seems to be the only formation for at least several miles up the valley. In this are the copper bearing veins, six or eight inches thick, of quartz, containing iron and copper-pyrites, a little zincblende, and some calcspar in cavities. The mine had only been opened a short distance.

Near the house there is a warm spring, with a temperature of 48° C., rising in the argillaceous rock.

June 4th. Leaving Kakumi, in the afternoon, we rode about three miles to the fishing village of Wosatzube.

Just east of the village is a promontory formed by an outcrop of beds of black hornstone.

Fig. 11

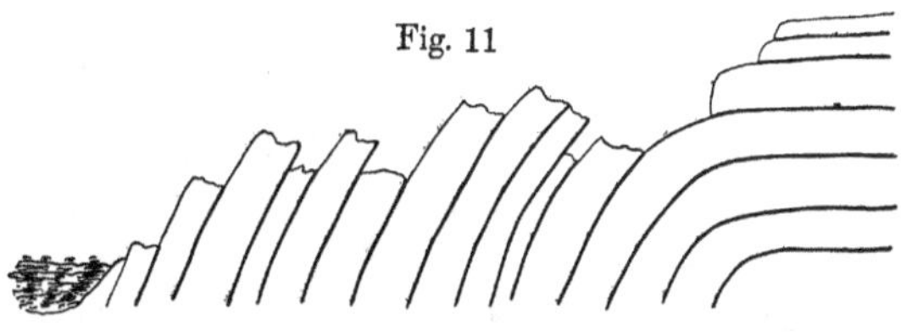

Hornstone Strata. Cape Wosatzube.

This rock is stratified in well-defined layers from a few inches to several feet in thickness. It has a velvety-black color, more rarely with lighter shades, breaks with conchoidal fracture, and shows, when wetted, a lamellar structure the layers of which are thin as paper, of black and dark-gray shades. In places it is slightly brecciated, the interstices being filled with opalescent chalcedony in layers of infiltration.

I may add that the Japanese mining officials who accompanied us, stated that a similar rock occurs in close connection with the coal beds on the eastern coast of Yesso. The trend of the strata at Wosatzube is N. 40 W., the general dip being northeasterly.

Off the point just described is a spring which bubbles up from the bottom, very strongly at low water, and quite visibly at high tide.

June 5th. The country east of Wosatzube being impassible for horses, we embarked in a boat propelled by sixteen rowers, and after a voyage of between three and four hours reached the fishing village of Totohoke. The scenery was very grand, as the coast is here formed by a wall several hundred feet high, and washed by the sea at its base. Innumerable waterfalls, some of them very high, and all beautiful, were seen at the heads of ravines, or falling like veils over the high coast bluffs. These cascades occur along the entire Japanese coast, and the early navigator Vriess mentions them at almost every step in his narrative.

The rock forming this coast wall seems to be volcanic tufa-conglomerate, with lava dykes. On examining the rock of the bluff west of Totohoke, it was found to be indistinctly stratified and made up of round and angular fragments of trachytic lava inclosed in a gray matrix more or less hard, with earthy fracture, and containing perfect crystals of hornblende and altered felspar, with scattered grains of quartz. The rock often presented in the fresh fracture all the appearance of an earthy lava, its detrital origin being most apparent on the weathered surface. The stratification dips northward toward the sea.

Totohoke lies at the foot of the volcano Esan.

June 6th. We ascended on horseback to the crater of Esan volcano, which forms the eastern point of the peninsula.

This, also, is a solfatara, its latest eruptions, of which there is no record, having been confined to flows of sulphurous mud. No pumice was seen, and the fragments of rock that formed the ejecta were of the same character as the walls of the crater, excepting some blocks that seemed to be pieces of the white quartz porphyry found at Kakumi, which had been torn from the interior of the mountain.

The crater, which seemed to be larger than that of the Sawaradake, is divided unequally by a high ridge of detritus. The walls, where observed in our passing examination, were found to be so altered by the constant action of acid vapors, as to render the character of the original rock very obscure, but I thought myself able to trace a similarity, through a series of specimens, between this and the more common ejected blocks. These latter consist of a dark gray cellular lava of porphyroidal texture. The crystals of felspar, which are numerous, are changed to a white earth, isolated specimens still retaining numerous crystals of hornblende; but the most characteristic feature is the abundance of quartz. This last mineral is present in well-defined, double pyramid crystals and in grains one-eighth to one-third of an inch in diameter. The grains are both limpid and milky white, and opalescent. They are highly fractured, and often present the appearance of having contracted and cracked in passing from a gelatinous to a hardened condition. There is often a strong resemblance between these rocks and the fragments inclosed in the tufa-conglomerate of Totohoke.

The walls of the crater are rapidly disintegrating and falling, to be converted into clay impregnated with sulphur, alum, and other salts. Everywhere the scene is one of ruin. Here is visible on a grand scale the decomposing action of sulphurous acid and steam, the effects of which we see in the altered trachytic rocks of Hungary, and still progressing on a small scale in the Neapolitan solfatara. Nowhere have I seen so well exhibited the levelling power of nature when she brings into action her more active agents.

Steam surrounds us, issuing in jets from fissures on the sides of the crater, and rising slowly, as smoke from a smouldering fire, out of the taluses of debris. But the main vents are small, mud craters or geysers. Those which we visited were in the centre of one of the divisions of the crater. They were springs or pits, each covered by a great vault of hardened mud, like an immense bubble or an inverted bowl, from ten to twenty-five feet high, the sides and roof from six inches to two feet thick.

These quake with the constant reverberation of the struggling steam and mud, which last, judging from the sound, must rise to near the surface. The inner surfaces of these vaults are lined with sulphur in massive layers, in crystals, and often in long stalactites, and the vapor is highly charged with sulphuretted hydrogen.

While we were here drops of scalding mud were incessantly thrown out, but regular mud-flows appear to be very rare.

The superintendent of the sulphur works informed me that when new vents open, mud and large blocks of rock are thrown out with much violence. Such blocks cover the interior of the crater, and have been already mentioned; they are frequently almost entirely decomposed by the action of the gases.

From an extinct vent I traced a stream of mud, following the bed of a gully, for several hundred yards. It is hard, compact, and filled with small crystalline needles of sulphur, the longer direction of which was found to be invariably at right angles to the nearest surface, by which either the heat or moisture, or both, escaped. These crystals occur equally distributed throughout the mass the whole length of the stream, and produce, on a small scale, a tendency to columnar structure. They cannot, considering their position, have been crystallized until the mud was quiescent and hardening, and as the solidification depended on the escape of the moisture that rendered it fluid, it forms, I think, a good illustration of the fact that columnar structure is not necessarily a result of cooling, but rather of the escape of the "vehicle of fluidity," whether this be heat or water, or, as here, both combined.

The stream in question appears to be the result of a single flow filling the inequalities in the bottom of the gully, and is in places several feet deep.

The government has large sulphur works on this mountain, with which the production of alum was formerly combined. The material used, from which the sulphur is extracted, is the debris formed by the ever-falling walls of the crater, and which is said to contain from 25 to 50, and even 60, per cent. of the mineral, in layers and impregnated through the mass.

Without further preparation than being broken with the hammer, this raw material is put into three iron pots over a fire. Each of these vessels is composed of two parts, a cylinder and a hemispherical bottom or pot on which it stands, the whole being about two and a half feet deep and two feet in diameter. After melting, the impurities seem to settle to the bottom, and the top is ladled out into shal-

Fig. 12

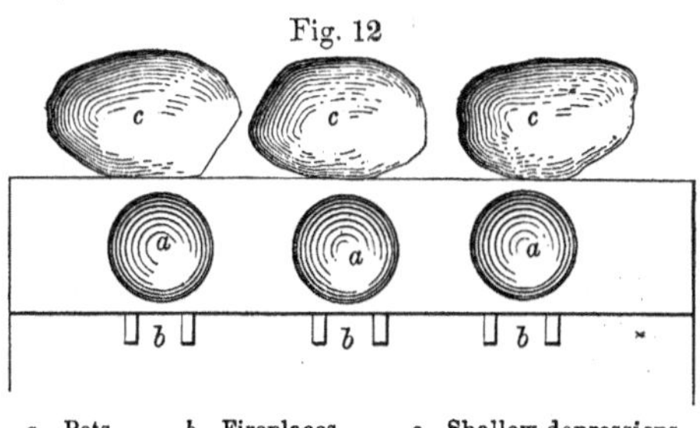

a Pots. *b* Fireplaces. *c* Shallow depressions.

low depressions in the ground. When this is cooled, it is a hardened mud filled with crystals of sulphur in needles, their longer axes at a right angle to the surface of the cooled mass, and the whole product differs from the mud described above, as having flowed from a vent, only in that the artificial product is richer in sulphur. In this instance the "vehicle of fluidity" was undoubtedly heat acting through melted sulphur.

This first rough product is remelted in similar pots, and then filtered through sacks, at first allowing the liquid sulphur to pass, by its own weight, and finally squeezing it gently under a lever. From these filters it falls into tubs the shape of which it retains on cooling. The blocks thus obtained are broken, and the cooling surface, to the depth of two inches, being of a dark color, and, perhaps, less pure, is remelted to obtain yellow sulphur; the interior of the blocks is yellow and highly crystalline.

The produce at the time of our visit was about 5,600 lbs. daily. The officials stated in round numbers that, everything included, the cost of producing 32,000 lbs. was about 80 rios, or $103, the same quantity bringing about $385 at the Hakodade market.

The iron pots cost for the top pieces $2 66 each; for the bottoms $6 60. The bottoms last from 30 to 60 days.

Continuing our journey we descended the western slope of the mountain to Nitanai, on the sea-shore.

June 7th. Leaving Nitanai, we rode along the sea-shore to Kobi. Near Nitanai we passed the outcrop of a bed of white infusorial earth raised several yards above the sea. The reader is referred to Mr. A. M. Edwards' Letter (App. No. 3) for the highly interesting results of his examination of this material under the microscope. Mr. Edwards has discovered a close resemblance between the organisms contained in this deposit, and those of the stratum under Richmond and Petersburg, Va.; and a still greater similarity to those of the extensive deposit along the California coast, the resemblance in the latter instance extending even to identity of species among the *Diatomaceæ*.

At Kobi an attempt had been made to smelt the magnetic iron sand from the beach in a blast furnace of the foreign pattern. One of our party, Mr. Takeda, a Japanese officer of rank, who has done much to advance, in his country, the knowledge of military engineering and navigation, was commanded by the Imperial Government to construct a large furnace for smelting iron ore after the foreign method. Such a thing had never been seen by a Japanese, but without further plans or specifications than he found in a Dutch work on chemistry, Mr. Takeda built a furnace about thirty feet high, after a very fine model, with cylinder blast moved by an excellent water wheel. Unfortunately, owing to the absence of all details on the subject in the only book he had, the blast obtained was only a fraction of that required, and the bricks used in the construction were not sufficiently refractory. Thus the affair was a failure after smelting a few hundred weight of iron. The incident, however, is an illustration of Japanese enterprise. I will add that the experiment was repeated by order of the Prince of Nambu, in order to work an excellent ore of magnetic iron on his property, and furnace after furnace built, from 20 to 30 feet high, until successful campaigns of several months' duration were obtained

At Kobi, besides the iron sand of the beach, there is an elevated, ancient beach, now from 50 to 100 feet above the sea, containing a bed of iron ore of a similar origin, the lower half cemented by oxidation to a solid mass, and changing to brown oxide, the upper portion less oxidized and retaining more of the original character.

How many deposits of iron ores may there not be that owe their formation to a similar cause, the destruction of ancient eruptive or metamorphic rocks, and the concentration of their grains of magnetic iron on the surf-washed beaches of former seas?

A few miles further on we came to the outcropping clay slates, which continue, as the tide-washed rock, as far as Shiwokubi (Cape Blunt). From this point on, as far as Oyasu, they are also exposed along the beach and form the hills inland, but are covered between the sea and the hills by the recent terrace deposit, which we have already seen bordering the Bay of Hakodade.

This slate is black and fissile, and is covered, near Shiwokubi by conformable strata of compact sandstone with interstratified seams of slate, and at Oyasu by a sandstone conglomerate containing fragments of the same older rock. These beds are more or less contorted, all the observed strikes of the uplift lying between W. and N. 15° W., averaging nearly N. W.

They are traversed by a great number of dykes of porphyry and greenstone, and by innumerable veins of quartz with pyrites of iron and, in places, of copper.

The porphyry is of the same white quartziferous variety as that at Kakumi, and the same description will do for both. The dykes are very sharply defined, from 10 to 50 feet thick, cutting the slates at all angles. The porphyry is in turn traversed by dykes of greenstone.

The quartz veins cut the slates at all angles, and vary in thickness from 2 to 12 feet. They abound in iron pyrites, one vein four feet thick being massive sulphuret. Some of them were traced between one and two miles inland, the pyrites changing to oxide away from the sea-shore. An outcropping vein at Saidoma showed some very fair ore of copper pyrites associated with iron pyrites, zincblende, and a little scattered galena. The strike of these veins is generally between N. and E., and one of the smaller ones traverses a dyke of porphyry.

It was in one of these that we made the first blast ever fired in Japan.

Between Shiwokubi and Hakodade, a broad *mesa* separates the hills from the sea, rising gently to near the mountain, and then rapidly, and cut into by all the streams descending from the hills. It is covered with a dense growth of weeds but no trees, the latter being confined, along this part of the straits of Tsungara, to the northern slopes of the hills.

At Yunogawa there is an outcrop of black clay slate in which rises a warm spring with a temperature of 38° C.

Entering Hakodade we finished the circuit of the peninsula.

The region thus encircled by our route is a high ridge apparently consisting, in the main, of the metamorphic rocks which have been described as occurring along the sea-shore, having a general northwesterly trend, accompanied by intrusive masses of greenstone and quartziferous porphyries. It is fringed on its northern slope by volcanic tufa-conglomerates that rise, in places, to the lower summits of the crest, and on the southern edge by recent marine strata. I will add that coal is said to have been found in the hills near Mt. Esan.

Excursion to the West Coast.

August 5, 1862. This day and the following one our route was about the same as on the preceding journey, as far as Volcano bay, where, branching off, we stopped at Washinoki for the night.

August 7th. Leaving Washinoki, we found, just west of the village, an outcrop, visible at low tide, of the tufa-conglomerate. It contained fragments of pumice and spines of an echinoderm. The beds are tilted up, the strike being N. 5° W. and the dip easterly.

A little further on we came to an outcrop of nearly vertical beds of a gray argillite, containing a peculiar fossil, having the shape of flattened vermiform tubes and changed to calcite. This organism although indeterminable is characteristic for this argillite, and served to distinguish the rock even when highly metamorphosed at many points on our journey.

I will mention here that between the bay and the mountains west of it, a strip several miles broad is occupied by a recent deposit, similar to that bordering Hakodade bay, and receding in terraces from the water which it faces with a bluff 30 to 80, or more, feet high. This deposit generally hides all the older rocks.

Continuing our journey along the beach, we found the tufa-conglomerate again in place underlying the terrace deposit.

Passing Otoshibetz,[1] the beach is overhung by the terrace bluff, here from 60 to 80 feet high. This recent deposit is a horizontally stratified, sandy clay, abounding in marine shells, chiefly bivalves. Although most of the shells were too friable to be collected, many seemed to have retained a large part of their organic matter, and in several instances I found the dorsal ligament still elastic when wet.

At Yamukshinai, just back from the beach, between this and the bluff, there is a marsh some acres in extent, in which tepid springs deposit a mineral oil of the consistency of tar, which is used by some priests, in the neighborhood, both for burning and in making ink of the kind used throughout China and Japan.

Passing through a settlement of Ainos we reached Yurup.

August 8th. The terrace bluff recedes from the sea at Yurup, forming a bight which is occupied by a broad plain, often marshy, covered with a dense growth of reeds and weeds, twelve to fourteen feet high. Through this plain winds the large creek Yurup.

Crossing this stream we followed the beach to Shirarika. Here there is an outcrop on the beach of a black amygdaloid, containing small spherical cavities lined with a white, transparent, tabular zeolite, and veins and nodules of chalcedony.

Continuing our journey over a plain, now sandy, now marshy, which, at the height of 10 or 20 feet above the sea, forms a narrow belt between the beach and the bluff, we reached Kunnui. The terraces seen during this day were covered with a fine forest growth of deciduous trees and scattered tall pines.

Leaving the sea-shore at Kunnui, we ascended the creek of the same name to a low pass in the crest, which here forms the watershed between Volcano bay and the Japan sea.

---

[1] The termination *betz* and *nai* are Aino words signifying river and creek or brook.

The only formation seen was the terrace deposit, till near the divide, when an obscure green wacke was found in place, and near this a greenish-black amygdaloid. Large blocks of granite were also seen here, and this rock is probably in place near by.

Descending to the west we entered the valley of the Toshibetz, a large creek, navigable with small, flat boats, and soon reached the gold washings of Kunnui.

This part of the valley occupies a broad depression, perhaps 15 miles long by 7 broad, and raised several hundred feet above the sea. It has been filled with the recent terrace deposit, and subsequently eroded in part, after which an extensive deposit of auriferous gravels, etc., has taken place over at least a considerable part of the area.

In one of the side valleys the older rocks are exposed, and here the gold bearing drift was found resting, in different places, on an argillite similar to that seen at Washinoki, and containing the same vermiform fossils, in strata striking N. 85° W., and dipping 50° northerly, and on an amygdaloid similar to that on the divide. Not far from here the terrace deposit overhangs the creek in a high bluff. Out of the base of this precipice I obtained a number of well-preserved fossil shells. In the same bed were found Ostreæ, Pecten, Scalaria, Terebratula, Nuculina? Serpula? Corals, Bryozoa, and fragments of a thick shell with cross-fibrous structure. Some of the shells retained, at least in part, their organic matter and nacreous lustre, and one species of Pecten appeared to be identical with a species living in the adjacent seas.

At one end of this bluff is a large rock of the amygdaloid in place, which has been exposed by the erosion of the terrace deposit, and on it are incrustations of Serpulæ.

This amygdaloid contains masses of a green rock resembling jasper, in which are scattered flakes of native copper. Blocks of manganese (binoxide) in the immediate neighborhood seem also to have come from the amygdaloid.

The auriferous gravel occurs along both sides of the river in the form of a plain, which descending gently from the hills faces the stream with a bluff. The whole district appears to have been worked in former times, though when appears to be unknown. Broad and deep canals of considerable length were dug to bring water from up the creek, and a well arranged system of "ditch diggings" seems to have been carried on. All these workings are covered with a dense growth of trees, apparently not differing from the surrounding forest; some seen in the ditches being as much as eighteen inches in diameter. The method of washing the gold does not seem to have differed from that now used by the Japanese.

The principal rocks, that have contributed to form the auriferous drift, are varieties of granite, chloritic and micaceous schists, quartzites, and amygdaloid, with geodes of chalcedony from the last mentioned rock. Rolled fragments of binoxide of manganese are frequent also, perhaps derived from the amygdaloid. The concentrated sand of the washing is principally magnetic iron associated with zircon sand.

The manner of working the deposit is ingenious, and will be understood by referring to the annexed diagrams.

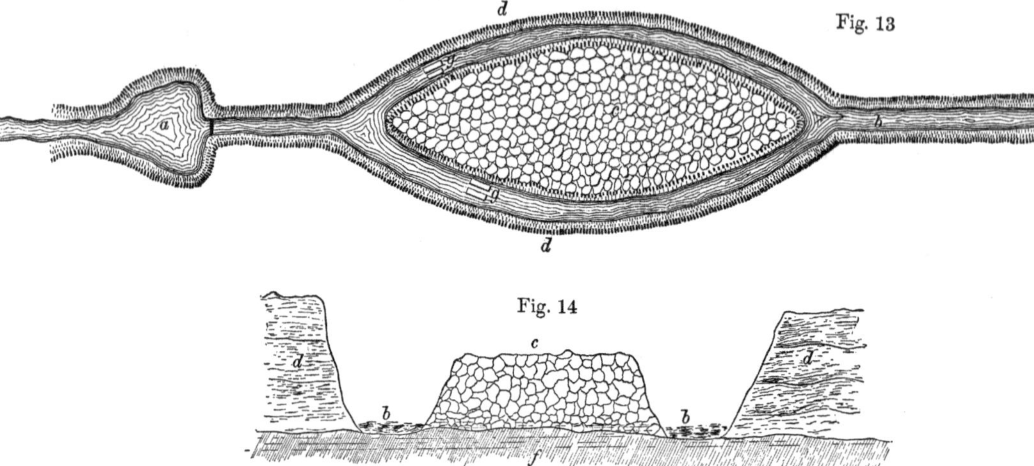

a. Reservoir. *b.* Sluice-ditch. *c.* Rubble of the drift. *d.* Aurif. drift. *e.* Creek. *f.* Bedrock. *g.* Mats.

At the place where I saw this process, the surface of the bed rock, in this case the marine terrace deposit, was sufficiently high above the creek to give a rapid fall in the sluice-ditch.

The bed of a rivulet is chosen for the work. A reservoir (*a*) is dug and dammed, and the bed of the rivulet (*b*) cleaned out and made regular. This done, the banks (*d*) are broken down into the stream where the force of the current concentrates the gravel, carrying off the sand and clay. The workmen then place themselves in pairs up and down the stream near and below the broken-down bank. Each man is provided with a coarse mat, about two feet long by one foot broad, which he places lengthwise in the stream, keeping it down with one foot on the lower end, at the same time partially stemming the current. He then hoes the gravel on to the mat, much of the old gravel going off below as fresh arrives from up stream.

At intervals the mat is carefully removed and washed out into a very shallow tray or batea (Fig. 15), a board about eighteen inches long by a foot broad, hollowed out, and having a circular depression near one end for the concentrated head. Of the black sand obtained on this board, the head containing the gold is saved.

Fig. 15

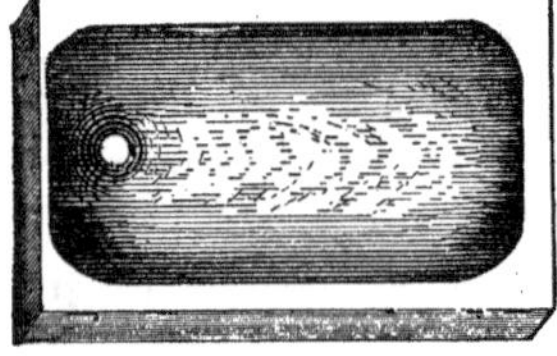

In this manner the gravel is pretty well exhausted of its gold, very little being obtained by the men farthest down the stream. The working progresses sideways, into the banks, and up stream, the current being kept near the banks as these recede from the centre of the stream. As the space between the banks widens, the coarser material that resists the force of the water is thrown up into a pile of loose masonry (*c*) which increases in length and breadth as the work advances.

Numerous remains of ancient workings, by this method, are found in the neighborhood.

Throughout this region the forest is dense; among the trees I noticed elms and a wild mulberry with black fruit. Fierce, large flies, of two kinds not seen on the sea-shore, swarm in these woods, covering horse and rider, and leaving bleeding wounds wherever they strike. The creek abounds in mountain trout and salmon.

August 14th. Returning to Kunnui on the sea-shore, we followed the beach to the village of Woshimanbe.

August 15th. At this village we left the bay to cross over to the west coast. For several miles the road lay over the terrace belt, here covered with drift. At the divide we found a broad, marshy tract through which a large creek winds on its way to the Japan sea. This stream we descended in a small flatboat.

The prevailing rock across this low part of the ridge was, so far as I could judge, an argillaceous deposit, apparently the same that forms the terraces.

The forest contained, chiefly, large beech, birch, and maple trees, with oaks and scattered firs, and the usual dense undergrowth of cane. The banks of the streams were lined with water willows. The creeks abound in trout, and the gravelly bottom is often nearly hidden by colonies of unio. As we approached the bay of Odaszu the country became more open, and leaving the creek we descended over two terraces of drift to the village of Odaszu on the sea.

The southern shore of this small bay is shallow and shelving, with a broad beach; but the eastern and western sides are rocky, the rocky bluffs descending into the sea, a feature common to all the west coast, so far as we followed it, and indeed to the shores of all the Japanese islands.

August 16th. Leaving Odaszu we continued our journey northward along the coast. Here, also, high terraces face the sea, but they are formed of the tufa-conglomerate formation, the level surface being due to a recent deposit of gravel and sand. This conglomerate is traversed near Odaszu by dykes of a dark gray rock, much weathered, containing crystals of a triclinic felspar, and opalescent chalcedony. The conglomerate at Isoya is traversed by dykes of an amorphous rock containing crystals of triclinic felspar.

Near Isoya there is a deposit consisting of beds of sandstone, argillaceous material, and volcanic ashes,[1] with fragments of pumice, and also of the argillite which has been mentioned as occurring at Washinoki and Kunnui with a vermiform fossil. The pieces of pumice contain beautiful double-pyramid crystals of quartz. This deposit is younger than the neighboring tufa-conglomerate, which had suffered much from erosion before the deposition of the beds in question. It continues northward till it abuts against a mass of volcanic rock, that forms the headland south of the mouth of the Shiribetz river. This stream rises nearly north of Cape Edomo, and flows westward through a fine, broad valley. All the gravel brought down by the river seemed to be trachytic detritus.

---

[1] For the interesting results of a microscopic examination of this material, see Mr. Edwards' Letter (spec. No. 11), Appendix 3.

Crossing the valley of the Shiribetz we came to the foot of the Raiden promontory, a bold headland presenting vertical cliffs toward the sea, and apparently made up of lava flows and tufa-conglomerate. In crossing this mountain we frequently found fragments of a black scoria with long-drawn cells.

After a laborious journey of several hours we descended into a deep and gloomy gorge containing a warm spring. Here again we found the same variety of white quartziferous porphyry that we had seen at Kakumi and elsewhere. It is impregnated with iron pyrites which in places is represented only by cubical cavities containing sulphur. The rock traversed by this porphyry is of a brecciated argillaceous character, resembling that at Kakumi. It is from this rock that the springs flow, with a temperature varying, in different ones, from 46° to 50° C. These rocks are exposed only in the bottom of the ravine, on either side of which they are covered by the volcanic formation.

August 17th. Rising from the ravine we continued our journey over the northern part of the Raiden, the outcrops here, as yesterday, being of a gray trachytic lava with a tendency to tabular structure. This continued till we descended at the creek Nibitzunai to a terrace that reaches many miles northward and eastward, low near the sea, but rising rapidly toward the mountains. Skirting this for a few miles we reached Iwanai.

August 18th. At Iwanai we left the sea and made an excursion to the volcano Iwaounobori[1] about thirteen miles inland.

The first five miles of the road lay over the terrace which, as we approached the mountains, rose very rapidly. During the first mile or two, after leaving the sea, the surface was covered with a dense growth of long-jointed grass, six or seven feet high, to which succeeded the usual forest of large maples, oaks, mountain and white-ash, beech, birch, fir, and scattered magnolias, filled in with an impenetrable undergrowth of cane eight to twelve, and even fifteen feet high. The road through this region, being deep with mud which was full of sharp pointed stumps of the cane, was one of the worst I have ever seen.

Entering the mountains we passed through a crateriform valley, once the bed of a lake, and, ascending to a pass in the hills beyond, we saw, beneath us, a beautiful little lake. On the other side of this rose the volcano, or rather solfatara, with its yellow, sulphur-coated cliffs. Here again the regular slopes and symmetrical outlines of an undisturbed cone are entirely wanting; the outer as well as the inner walls were rocky precipices, and the ruin seemed greater than at Esan. We reached the summit without much difficulty.

The present mountain is evidently only part of the skeleton of a former cone of large size. The predominating formation, from the spurs at the base to the summit, is a dark gray volcanic rock, showing in places a tendency to stratiform structure, and apparently of the trachytic family, the chief ingredient being crystals of a white felspar.[2] The former mantle seems to be still represented by fragmentary

---

[1] Japanese. *Iwaou*, sulphur; and *nobori*, a term for mountain, from *noboru*, to climb.

[2] With the exception of one specimen of rock, and a few minerals, the entire collection of rocks, shells, etc. from north of Odaszu, was lost by the wreck of a junk on the way to Hakodade.

remains of a stratified deposit seen here and there, about the base, and fragments of scoriæ were found in the neighborhood.

There are several small crateriform depressions at different points near the summit, filled to the level of the lip with sand and clay, and forming small plains surrounded by rocky sides. In one of the walls a compact black rock, either a dyke or the remnant of a lava flow, was observed.

The Iwaounobori is the central one of three volcanoes, which lie in a straight line running about N. N. W., S. S. E., and this is also the trend of a broad belt, within the limits of which the solfatara action is most developed, both across the summit and on the outer walls.

Throughout this belt the rock, wherever not covered by the products of decomposition, is found to be traversed by countless fissures, more or less filled with sulphur. Wherever the filling is incomplete, small jets of steam and gases are still seen to issue forth. Several trials, made by inserting a long chemist's thermometer as far as possible into different fissures, gave a constant temperature of 98° C.

The steam has a strong odor of both sulphurous acid and sulphuretted hydrogen. It has an acid reaction on litmus paper, which is especially strong when the condensed drops, that hang on the sulphur crystals in the cavities, are tested. Beautiful crystals of sulphur, a quarter of an inch long, were rapidly formed on the bulb of the thermometer.

Excepting at the steam vents, which are not more than from one to five inches in diameter, the fissures are closed up with sulphur at the surface, but by breaking away a few inches deep, cavities are exposed lined with a bristling mass of most beautiful straw-colored crystals of this mineral, made up of brilliant steep pyramids connected in the line of the longer axis. Unfortunately, they were too delicate to bear transportation.

On a precipitous part of the outer wall of the mountain, where a large mass of rock seemed recently to have fallen off, I saw an interesting exhibition of the action of the gases. The rock is seen to be traversed by a perfect network of sulphur veins (*a*) which seem to occupy the positions of the cracks common to all rock. The trachytic rock (*b*) is tolerably well preserved in the centre of the blocks, but toward the circumference it is more and more disintegrated, and has assumed the form of concentric layers, the outer shell being changed to a white earth. It seems not improbable that this condition may exist through a large part of the mountain, thus forming a great *stockwerk* of sulphur.

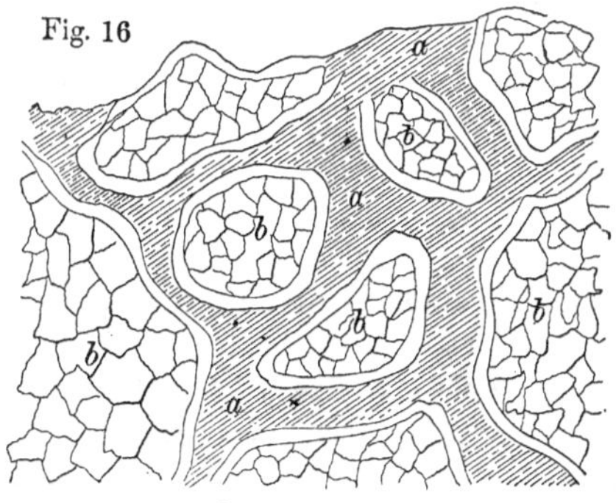

a. Sulphur. *b.* Rock.

The only way in which I can account for this structure is, by supposing that the disintegration of the rock, which formerly occupied the spaces now filled with sulphur, took place when the water, which now appears only as steam, stood at a higher level in the mountain, making it a mud volcano, like Esan, and exuding the products of decomposition as fast as formed. On the withdrawal of the water to a lower level the abandoned network of fissures was filled by the decomposition of sulphuretted hydrogen.

At another place, in the walls of one of the small craters near the summit, there is an instance that would seem to illustrate the action of the gases and steam without the presence of water as such. The black rock, already mentioned as occurring in the wall of one of the craters, is visible in different stages of alteration. In places it was observed to have the concentric structure assumed by many rocks during the first period of disintegration, and by which the polygonal form of the blocks, into which all bodies of rock are subdivided, is lost as each succeeding shell is removed. In this case the outer shell is white and earthy. Again the same rock was found altered to the centre of each block, the shape remaining, to a soft, pasty, white clay, quite tasteless. Often in the centre of a snowy white mass of this clay would lie a core, equally soft, but black, the line of separation between the colors being well marked. In places, where the alteration was in the first stage, an alum salt was found forming an efflorescence on the surface of this black rock, possibly as one of the first products from the decomposing felspar.

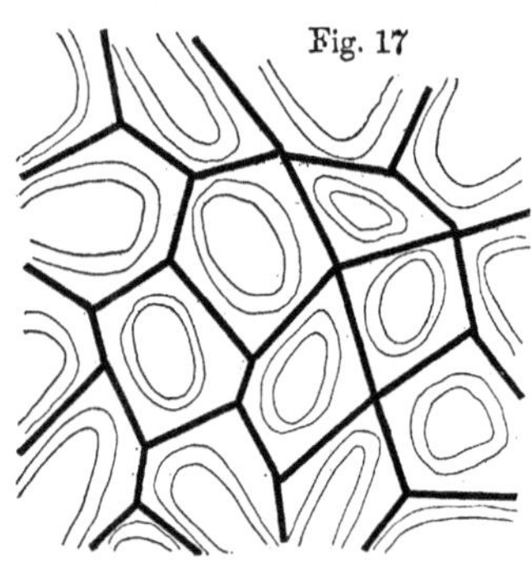
Fig. 17

An emerald-green soft mineral occurs incrusting, to the depth of a line or more, the walls of the gully where these phenomena were observed.

On the west side of the peak, in the valley which drains the craters, there was formerly a spring of chalybeate water, which has left quite a deposit of oxide of iron filled with the leaves of a cane, apparently of the same species that covers the surrounding country. At present there is no cane on this part of the mountain, although it grows within a few hundred yards of the spot. This space, which is bare of cane, abounds in Winter-green (Gaultheria) with white berries.

In close proximity to this deposit a white altered rock, filled with threads of sulphur, attests the former action of the gases in this spot which is now removed from the nearest field of activity.

From the summit of the Iwaounobori I counted fifteen mountains, all of which seemed to be of volcanic origin. Among these I include Esan, Sawaradake, and Oussu, all solfataras, which, from their ruined condition, I would not have recognized as volcanoes at this distance had I not known them to be such.

A few miles away to the S. S. E., beyond the broad valley of the river Shiribetz, rose a magnificent cone also called the Shiribetz. This cone is the most symmetrical of any that I have seen, not excepting the beautiful Fuziyama, the pride of the Empire. Of its height I had no means of judging, but I thought it could not be less than 6000 feet. It rises from a broad plain, at least the slopes visible to us merged gently into the sweeping cross curves of the valley of the Shiribetz river. The unbroken surface of its sides was covered from base to summit with vegetation, either forest or cane, which appeared to us in the distance like a mantle of green velvet. Many other well-shaped cones were visible in the distance.

Just N. N. W. of the Iwaounobori there is a cone somewhat lower than the peak of the solfatara, with a well preserved crater, so near that it seems to be partly within the circumference of the foot-slope of the Iwaou mountain. As I have said before, it is in a line with its neighbor and the Shiribetz, and this direction is repeated in the zone of the solfatara activity on the Iwaou mountain, a coincidence that would seem to point to a fissure connection between the three peaks.

The government has sulphur works on this mountain, in which fourteen caldrons are kept at work. The production is about 64,000 pounds per month, costing for—

Labor of all kinds and for fuel per month	$74 50
Rice for workmen	41 00
Salt and miso for workmen	4 00
Straw sandals for workmen	6 50
Transportation by horse to Iwanai	57 25
Total for 64,000 pounds	$183 25

August 20th. We returned to Iwanai.

August 21st. Continuing our journey northward, we rode along the beach to the mouth of the Shiribuka creek, where the coast line, turning off to the northwest, marks the southern shore of the peninsula south of Strogonof bay. Following this shore we left the terrace plain of Iwanai bay. During the rest of the day we saw only the tufa-conglomerate formation, which, traversed by numerous dykes of volcanic rock, faces the sea in bold bluffs, to pass which we were at last compelled to take a boat to carry us to Ousubetz, a small fishing village.

The volcanic conglomerate of this region extends some distance inland, and consists almost entirely of more or less rounded fragments of black lava filled with green-coated cells.

August 22d. Leaving the sea we made a short excursion up the bed of a creek, the Kaiyanobetz. About one mile from the shore a gray sandstone was found exposed for a short distance beneath the volcanic conglomerate, and about one mile and a half further we found in the bed of a rivulet the following strata, the order reading from younger to older.[1]

1. Fine-grained argillaceous rock with fossil plants.
2. Coarse sandstone.
3. Clay shale with *Equisetaceæ*.
4. Coarse sandstone.
5. Three seams of bituminous coal alternating with thin beds of clay, the principal seam having about four feet of good coal.

The strike of these beds was N. 30° E., the dip being 50° to N. 60° W.

In a neighboring ravine a white silicious rock was observed, apparently older than the coal, and made up of minute layers, the whole being hard, and having somewhat the appearance of a semi-opal.

---

[1] Except a small specimen of coal which was brought away by one of the Japanese officers, all the collections from this region were lost in the wreck mentioned above.

Retracing our steps to Ousubetz we embarked in a boat propelled by eight oars men, four scullers, and a large sail, and soon reached Iwanai.

August 25th. Leaving Iwanai we went by boat to Isoya, passing close under the rocky cliffs of the Raiden. The northern part of this mountain is formed of the volcanic tufa-conglomerate covered by a great bed, or perhaps several flows, of lava, often exhibiting columnar structure. In places beds of lava seemed to be interstratified with the conglomerate.

At about half the distance between the northern and southern sides of this highland, a large amphitheatre or crateriform valley opens towards the sea. South of this the cliffs, less high, consist of the conglomerate, and in the perpendicular walls are visible many small but regular dykes with transverse columnar structure, and in places dislocated by faults. The conglomerate strata have a considerable southwesterly dip, and as we approach the southern flank of the Raiden, near the village of Hamajimé, they disappear under the sea. Overlying this formation and forming the mountain above, is a gray volcanic rock, possessing a tabular structure, which gives it often a stratiform appearance near the bottom, but in the upper half of its thickness the plates curve irregularly upwards, presenting their edges towards the upper surface of the bed.

This mountain is a high, flat ridge, running nearly east and west, between the valleys of the Shiribetz and the Shiribuka rivers, and on it is the Iwaou nobori, and at least one more volcano.

August 27th. Leaving Isoya, we rode around the head of Odaszu bay to Sutzu. On this side of the bay we met again terraces of conglomerate, covered with loose sand and gravel, corresponding to those mentioned as occurring on the opposite side.

Before reaching Sutzu the conglomerate formation was found to be succeeded, for a short distance, by a gray eruptive rock, apparently a trachytic porphyry. The conglomerate in this region consists, almost entirely, of rounded fragments of a compact black rock, almost a pitchstone, containing crystals of white triclinic felspar.

August 28th. Leaving Sutzu we rode westward, over the lower of the two terraces that rise between the sea and the hills. The highlands are wooded with small trees, but on the terraces there is generally only a heavy growth of weeds and joint-grass, often from six to ten feet high. Leaving the sea-shore, we crossed the promontory to its western flank, travelling over the conglomerate, upon which was seen a loose deposit of sand and gravel closely resembling the auriferous deposit of Kunnui. In one place I observed an outcrop of the argillaceous rock, with the peculiar vermiform fossil, seen at Kunnui, Washinoki, etc.

At Achase the tufa-conglomerate dips inland, and beneath it there is an apparently conformable bed of fine-grained, brown sandstone, easily scratched with the knife, and seemingly of the same origin as the conglomerate.

A few miles further southward we reached Shimakomaki. Here the semi-vitreous character of the pebbles that compose the conglomerate is better developed than usual, although a black amorphous base was found to be generally prevalent, in these fragments, in the tufa-conglomerates of the west coast. Here the base of the rock is jet black, opaque, with the lustre of pitch, and imperfect conchoidal fracture. Fragments break off with a very hackly surface. The structure varies from slightly cellular to scoriaceous, the cells being lined with a light greenish or bluish film. It contains thin crystals of white, glassy felspar, the number of which seems to be in an inverse ratio to that of the cells. The felspar is, at least in part, a triclinic variety.

The Tomari creek, which enters the sea near Shimakomaki, brings down among its rubble, diorite, granular limestone containing nephrite, clay schist, and varieties of quartz and jasper. This stream rises in the hills that have furnished, in part at least, the auriferous gravels of Kunnui, and it is probable that similar deposits occur also in the valley of the Tomari.

August 29th. Embarking in a large boat we sailed close under the lofty cliffs of a grandly picturesque, but dangerous coast, as far as Setanai.

The volcanic conglomerate exists as the principal formation of the coast, between Shimakomaki and Setanai. At Cape Shiraita the thickness of the conglomerate, above the sea, is between 100 and 200 feet; above this is a bed, perhaps 150 feet thick, apparently of a looser material, with many white fragments scattered through it; and, finally, covering this, for a distance of one or two miles, is a bed of lava, 150 to 200 feet thick.

From this point to Cape Moteta the cliffs are entirely of the volcanic conglomerate, of which a lower bed is sometimes visible, with white fragments, those of the upper beds being dark brown or black.

At Cape Moteta the volcanic conglomerate, occupying the lower part of the cliffs to the height of between 100 and 200 feet above the sea, is covered by a thick bed of columnar lava. Near this point a broad dyke rises through the conglomerate to the overlying lava bed, but it was impossible to determine, at a distance, the relative ages of the latter and the dyke.

Numerous dykes traverse the conglomerate between Cape Moteta and Setanai. At Abura the latter approaches sandstone in texture; at one place it was seen to pass abruptly into a white deposit, probaby a pumiceous tufa.

South of Abura the conglomerate is covered by a lava bed, and this by white, apparently tufaceous, strata.

Several miles north of Setanai a thick bed of columnar lava is visible, high up the face of the cliff, lying between two members of the neptuno-volcanic formation, and dipping gently toward the south. Before reaching Setanai a thick flow of lava, beautifully columnar and probably the continuation of the bed just mentioned, occupies the lower half or more of the cliff, while needles of the same rock rising high out of the sea form picturesque islands.

This rock is a dark brown, much weathered, cellular lava. The cells are coated with a soft, brittle mineral, dark green in the fracture, and light bluish-green on the surface; and being flattened and parallel, with their planes at right angles to the axes of the columns, they give to the rock a slaty structure. Overlying this lava bed there are strata of tufa-conglomerate, made up mostly of fragments of cellular and scoriaceous volcanic products.

Just south of Setanai the Toshibetz—here several hundred feet broad—the river, on which lie the gold washings of Kunnui, empties into the sea—its valley, here several miles broad, being the first break, of any size, in the uninterrupted line of cliffs south of the Bay of Odaszu.

August 30th. Continuing our journey southward we followed the beach, separated here by high sand hills from the flats of the Toshibetz, till Futoro.

Just before reaching this village we left the valley and came under a bluff of trachytic or phonolithic lava, with a tendency to slaty structure. It has a light gray base, with semi-vitreous lustre, and is cellular—the cavities being very irregular in shape and lined with a grayish-blue botryoidal mineral. It contains numerous crystals of a glassy triclinic felspar.

At Futoro the volcanic conglomerate reappears as a red and brown tufa, with fragments of the lava just described and other varieties that show a regular transition from this lava into a black amorphous kind closely resembling that mentioned as forming dykes at Isoya. The strata of this neptuno-volcanic formation strike nearly N. and dip to E. about 20°, and the cleavage planes of the lava-bed described above dip in the same direction. This lava flow seems to be at least 250 or 300 feet thick. Just south of Futoro the contact between the lava and conglomerate was observed. The former rock at a little distance from the contact was found to be fresh, generally free from cells, and had a light gray compact base, abounding in crystals of triclinic, glassy felspar, with here and there a crystal of hornblende. Its appearance reminded me strongly of some non-quartziferous felsitic porphyries. Near the contact it became more earthy, and assumed the appearance of the base of the conglomerate, from which it was here distinguishable only by the crystals of felspar. The whole appearance of the contact seemed to indicate that the lava had flowed over the surface of the older deposit before this had become compacted.

August 31st. From Futoro we went by boat to Oöuta. Not far from Futoro the volcanic formations were seen to rest upon a granite or syenite, which, a little further south, abuts, with a vertical line of contact, against a compact black, aphanitic rock. This last was seen, in the face of a rock rising from the sea, to be traversed by veins of granite which, just south of this, was found to form the high cliffs till near Oöuta.

At Nichinbe, about three miles north of Oöuta, the prevailing rock was found to be a very beautiful syenitic granite, composed of greenish-white triclinic felspar, brilliant hornblende, black mica, and quartz. It is traversed by a dyke of a green, micro-crystalline rock, containing felspar and hornblende.

At Oöuta there is an extensive development of metamorphic rocks, consisting of a fine-grained granulite of even texture, and a conglomerate-breccia of argillaceous rocks. The only traces of a trend observable was in the vertical plane of contact between these two rocks, and this lay N. and S. South of Oöuta syenite reappears, and is shown to be younger than the granulite by the numerous fragments it incloses of the last-mentioned rock.

The granulite is cut by dykes of an aphanitic rock similar to that observed south of Futoro, and which we have seen to be traversed by veins of granite. Finally, the conglomerate-breccia incloses fragments of amygdaloid resembling a variety found in the auriferous gravel of Kunnui, and containing nodules of chalcedony surrounded by a soft green mineral resembling delessite.

The relative ages of the metamorphic and intrusive rocks of this region appear to be as follows, reading from younger to older:—

1. Greenstone of Nichinbe; dyke in syenitic granite.
2. Syenitic granite.
3. Aphanitic rock.
4. Metamorphic conglomerate and granulite of Oöuta.
5. Amygdaloid.

September 1st. Continuing the journey by boat we reached Kudo—the syenitic granite forming high hills along the sea as far as Ouenkoto, near Kudo.

At Kudo other metamorphic strata were observed, consisting of black and rose-colored quartz-schist, clay slate in thin beds, and a dark brown, micro-crystalline rock, apparently felspar and hornblende. These strata are folded and refolded, and the stratification being well preserved, they presented the finest example of plication I had ever seen. The general trend of the folding seemed to be about E., but there was too much irregularity in this respect to make sure of the direction; further south the trend appeared more regularly N. W. and the dip N. E.

The beds are traversed by a dyke of a porphyritic rock containing crystals of green and greenish-white triclinic felspar and of hornblende, in a grayish purple base.

A cold spring of chalybeate and carbonated water rises on the beach from the quartzite.

September 2d. Riding along the sea-shore, a few miles, we reached the penal establishment of Ousubetz, at the mouth of a creek of the same name.

Ascending this stream, which is a wild mountain torrent contained, near the sea, between cliffs of the volcanic conglomerate, we came upon an amygdaloidal rock, and beyond this a chloritic granite containing, besides quartz and chlorite, white orthoclase and a light green triclinic felspar. In this granite there is a broad belt, apparently a dyke, of a claystone-porphyry, a yellowish rock with a rough, earthy base free from visible quartz, and from which the crystals of felspar have disappeared, leaving only their cavities. From this porphyry issue several springs, which showed in different instances temperatures of 55°, 58°, and 58½° C.

These springs have formed deposits, of carbonate of lime and brown oxide of iron, which are more or less cavernous, and are the abode of a great number of snakes, which, attracted by the perpetual warmth, and being respected by the natives as the deities of the place, live unharmed. The cast-off skins of these reptiles flutter, like streamers, from every hole and neighboring bush.

Beyond the chloritic granite we found again the amygdaloid which, under various forms, extended as far inland as our excursion continued, about one mile beyond the chloritic granite.

In one of the side ravines a bluish-white, highly silicious rock, with conchoidal fracture and impregnated with minute cubes of iron pyrites, was observed in contact with the amygdaloidal rock.

This amygdaloid is very variable in character, in places brecciated, in others massive—the base being generally dark reddish-brown, and containing nodules of calcite and a green, soft clayey mineral, with here and there one of quartz. Fragments of a green serpentinoidal rock, which seemed to be a variety of the amygdaloid, occur in the creek.

September 4th. Descending to the sea we rode southward along the shore, under cliffs of the volcanic conglomerate, as far as the large village of Kumaishi.

September 5th. Leaving Kumaishi we followed the beach southward. From the village south the shore bluff is formed by a vertical cliff of white pumice-tufa, sufficiently hard to permit the making of steps in it. It is in thick beds having a southerly dip. South of Hiratanai this pumice-tufa is covered by the usual tufa-conglomerate.

A short distance east of Hiratanai a flow of amorphous lava, resembling that which occurs in fragments in the conglomerate of Isoya and Futoro, flows over the face of the bluff—the erosion of the conglomerate having progressed to nearly its present condition before the flow. A conical hill with a crateriform depression, lying several miles inland, was observed from the beach, and was possibly the source of the stream.

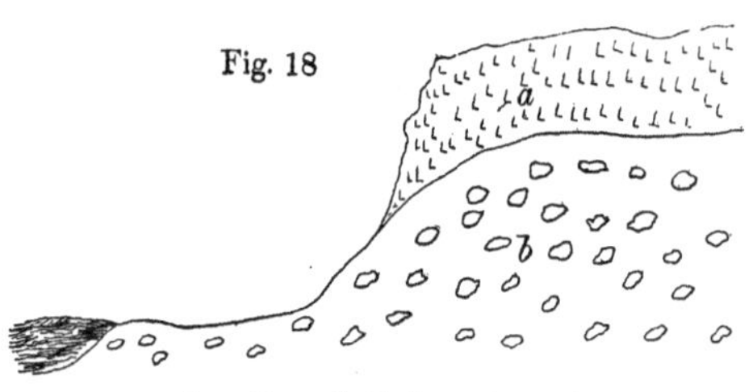

a. Lava flow. *b*. Tufa-conglomerate.

Beyond this point, as far as Tomarigawa, another bed of pumice-tufa, overlying the conglomerate, forms the bluff-rock and the skeleton of the terraces that extend several miles inland.

At Tomarigawa we left the sea-shore and entered the mountains, and ascending to the watershed between the Japan sea and Volcano bay, we descended the eastern slope to the mines of Yurup.

Our road, during this distance, lay, all the way, over the volcanic tufa-conglomerate formation, which extends entirely across this part of the island, and forms the ridge at a height of perhaps 2,000 feet.

This deposit is cut up by deep valleys with steep sides. In these I noticed outcrops, beneath the conglomerate, of granite, two or three miles from the sea, and, further eastward, of the argillaceous rock with vermiform fossils already mentioned several times.

The lead mines of Yurup are in the valley system of the river of the same name. Here a widely extended erosion has removed the volcanic conglomerate, for a considerable distance, exposing a very extensive development of a black metamorphosed argillite, which was found to contain the vermiform fossils so often mentioned in the previous pages. The strata are tilted up, often almost vertical, and are frequently connected with broad bands, apparently dykes, of greenstone. The lead-bearing veins occur in both these rocks. The vein-mass consists of quartz, carbonate of manganese, calcite, and, in one vein, crystals of barytes. Besides these minerals the galena is associated with zincblende, and pyrites of iron and copper.

The veins vary from two to eighteen inches in thickness, being more regular in the greenstone where, also, the gangue is chiefly quartz, and often existing as a zone, several feet broad, of parallel threads, in the argillaceous rock.

The mines have been worked several years and a considerable area explored, but like those at Ichinowatari they are very poor—the highest production ever attained being about four tons per month, and at the time of my visit it was only about one and three-quarter tons.

The processes of separation and smelting are the same as at Ichinowatari. The laborers are furnished, at the expense of the mine, with rice and *miso*, a vegetable substance used for soup. I have added a schedule of the daily expenses, more as a curiosity, and as illustrating the cost of labor, than for any other reason.

Daily Expenses of the Yurup Lead Mines.

Accountant clerk	$ 05
Head miner	07
Twenty-five miners, at 5 cts.	1 25
Eighteen coolies, at 4 cts.	72
Thirteen women ore dressers and washers, at 2 to 6 cents.	45
Daily consumption of iron	12
" " steel	04
" " mats and ropes	06
Total	$2 76

The working time is eight hours daily. The miners receive tasks, for all work over which they are paid extra. The task when working in the hardest rock, here a greenstone, is $\frac{3}{10}$ of one foot in five days, per man. In very soft rock five feet in five days, per man. The average is about one and one-half feet. The above measures refer to galleries five feet high and three broad. The miners are required to hew the walls as smoothly, and square the angles as accurately as was the custom in Germany before the use of gunpowder.

A woman's daily task is to pulverize about 160 pounds of ore.

One thousand pounds of roughly-sorted ore yields 67 pounds of *schlich*, from which 45 pounds of metallic lead are obtained.

The charcoal for smelting is produced in vaulted furnaces, which receive daily 64 cubic feet of split wood.

Both cold and warm chalybeate springs rise in the metamorphic argillite; the warm one, having the temperature of 46° C., is used in winter for washing the ore.

At this place we introduced the use of gunpowder in mining—its application to that purpose being entirely unknown throughout Eastern Asia. We met with the same objection here that was used, centuries ago, against its introduction into the German mines, the fear that the mountain would fall in. One blast, however, allayed this fear, and the miners adopted it enthusiastically thenceforth.

September 11th. Leaving Yurup we descended the valley to the sea. At the distance of about one mile from the mines we came again to the volcanic conglomerate. This formation is here similar in character to that seen between the Japan sea and the mines, but differs from that generally met with along the sea-shore. It has undergone so much alteration that it is often difficult to draw the line between the inclosing mass and the fragments. These latter are of a dark, cellular rock with amorphous base, containing abundant crystals of hornblende and felspar. The cementing material is a more or less yellowish mineral, with the lustre of wax, and easily scratched with the knife. This mass also abounds in crystals of hornblende and felspar, and is cellular in the same manner as the inclosed fragments. Specimens show a transition from one to the other, and this is especially observable around the cells in the fragments. The general color of the rock is dirty yellow. If this be not a true palagonite tufa it must be closely related to it.

The strata of this formation dip gently, on the western slope, towards the Japan sea, and on the eastern slope, towards Volcano bay. They consist of two principal members, the lower, a fine-grained, soft tufa with black mica and fragments of nearly decomposed pumice; and the palagonite tufa, if I may call it such, as the upper member.

At about half way between the mines and the sea we came again upon the argillaceous rock of the mines, containing the same characteristic fossil, but unmetamorphosed, and presenting itself as a soft gray argillaceous shale.

At the village of Yurup, on Volcano bay, we came into the road followed in going north, and completed the circuit of this itinerary.

Without attempting, in the absence of necessary data, to determine more closely the ages of the rocks referred to in the preceding pages, they may be generally classed as follows:—

I. Older metamorphic.

II. Pluto-neptunian.

III. Recent, including the marine terrace deposits.

IV. Eruptive, of all ages.

The first of these divisions contains all the sedimentary rocks that were observed to be older than the volcanic tufa-conglomerate formation. They are rocks that vary widely in character, and perhaps as widely in age. Forming the skeleton, of at least the southern part of Yesso, they are almost everywhere concealed by the younger deposits.

The most highly metamorphosed and perhaps the oldest strata observed are the granulite and conglomerate-breccia beds of Oöuta, on the west coast. These last are made up of older argillaceous and amygdaloidal rocks, but are also older than three varieties of eruptive rocks—aphanitic trap, syenitic granite, and a greenstone trap, apparently diorite.

The greatest part of the southeast peninsula, lying between Volcano bay and the Straits of Tsungara, is formed of fissile clay slates with subordinated beds of sandstone and conglomerates, the uplift trending nearly as the peninsula, about N. W. by W. These strata are traversed by frequent dykes of the characteristic white quartziferous porphyry, and varieties of greenstone, the latter being younger than the porphyry.

At Wosatzube, on the northern side of the peninsula, there are beds of silicious schist, having also a northwesterly trend, and strata of a similar character occur at Kudo, on the west coast, associated with subordinated clay slate and beds of a hornblende-felspar rock. Here also the mean trend of the highly contorted beds is between W. and N.

The remaining older rocks of this part of the island belong to the Ichinowatari series, and the argillite beds containing the obscure vermiform fossil, so often mentioned. The Ichinowatari series are black and gray metamorphosed argillacecus rocks, associated with older or younger shale containing calamites of unknown age, and with greenstone; and they are characterized by metalliferous veins occurring at least in both the argillaceous rocks and in the greenstone.

The argillite beds we find at many points, throughout the region included in the above itineraries, occurring in places either as a compact gray rock or as a shale, while at Yurup it is metamorphosed to a compact black rock, tilted almost to perpendicularity. Between Tomarigawa, on the west coast, and Yurup, on Volcano bay, it is found, excepting in one locality, to be the predominating rock wherever the ravines have cut through to the bottom of the volcanic tufa-conglomerate strata. The rocks in question have, in common with the Ichinowatari series, their argillaceous character, their association with dykes and great masses of greenstone and an identity of character in the metalliferous veins of the two localities, both as regards the association of minerals in these and also as regards some peculiarities in the condition of the greenstone near these veins.

Finally we have seen, beyond Iwanai, near Ousubetz (north), a coal-bearing series of more or less metamorphosed rocks, containing fossil *Equiseta.*

We find, in the auriferous gravel of Kunnui, representatives of another class of metamorphic rocks in the chloritic and micaceous schists, etc., which are probably the source of the gold, and evidently exist *in sitû* in the ridge between that place and the Japan sea.

The enumerated strata form, so far as my observation extended, the skeleton of Southern Yesso. The local strike of the coal-bearing rocks of the Ousubetz (north) is N. 30° E., being nearly at right angles to the N. W. trend of the peninsula on which they occur. All the other beds of the older rocks seem to have been affected chiefly by an uplift trending between N. and W., and to which that portion of the island lying between Esan volcano and the mouth of the Toshibetz, on the west coast, appears to owe its direction.

We come now to the pluto-neptunian beds, consisting of great masses, more or less stratified, of volcanic products in the form of tufas, sandstones, and coarser conglomerates and breccias.

This, by far the predominating formation, forms almost everywhere sloping plains or terraces between the mountains and the sea-shore, and extends, at least in places, entirely over the watersheds between Volcano bay and the Japan sea, forming peaks, as the Obokodake, several thousand feet high.

The petrographical character of these beds is very different, not only in their vertical, but also in their horizontal development. Along the west coast we find thick beds of a white pumice-tufa associated with conglomerates made up of fragments of a black compact rock, almost a pitchstone. Along the road from Tomarigawa to Volcano bay the lowest beds observed were of a more clayey pumiceous tufa, and above these an immense development of a scoriaceous conglomerate breccia, altered in great part to a wacke and strongly resembling palagonite-tufa. Bordering the eastern end of the southeastern peninsula, we have seen the representative beds of this formation, but differing from those of the west coast in that the inclosed fragments have more the character of quartziferous trachytic porphyry, thus approaching closely in character to the wall rock of the Esan crater and its recent *ejecta*, as also to the rock of Hakodade peak.

The only traces of fossils observed in this formation, were some fragments of the spines of an Echinoderm found near Washinoki.

The presence of these deposits over so large an area, and the fact that they always contain beds of coarse material, points to a corresponding range of volcanic activity. The same is indicated in the numerous lava flows and dykes that are intimately associated with these beds.

They are probably of submarine origin, and since their formation the island has undergone many changes of level. A large part of Southern Yesso was under water during the deposition of these deposits; it seems to have been gradually elevated and submitted to littoral erosion, forming the different terraces, and then to have been partially submerged to receive the recent terrace clay deposits.

This recent terrace deposit exists as beds of clay, almost exclusively, along the southern slope of the southeastern peninsula, and bordering the western shore of Volcano bay, and in depressions inland from this, as in the valley of the Toshibetz. Along the west coast where the depth of water is great, and the coast precipitous, this deposit rarely exists as clay, and then only bordering deep indentations like the Bay of Odaszu; but it is perhaps partially represented by the gravelly covering of the volcanic conglomerate terraces. As has been already stated, this terrace-clay deposit abounds in the remains of recent Mollusks.

After the elevation of these recent terraces, and after the action of an extensive erosion, there were formed the auriferous gravels of Kunnui, and finally, more recent and still progressing, subaërial deposits, as the volcanic-ash beds around Comangadake.

Very little is known of the physical character of the rest of Yesso. Volcanic cones, extinct and active, seem to exist throughout the island. Coal occurs at several points on the east coast, and several ammonites and a piece of obsidian were shown to me by the Governor of Yesso, as coming from the Monbetz creek, on the northern coast.

The island receives an additional interest from being a point of intersection of three lines of upheaval, and evidently owes its remarkable shape to this fact.

The first of these lines is represented by the northwesterly trend, of that portion of the island extending from Esan volcano to the mouth of the Toshibetz, and this is also the trend of the uplifted metamorphic strata; indeed the southeastern peninsula seems to be an anticlinal axis, the dip of the beds being on both sides, along the coast, toward the sea. This is also the trend of the peninsula south of Strogonoff bay, and of the northern coast line.

The second line is that extending from the headland of Matzmai, northeast through the longer axis of the island and of the Kurile chain to Kamschatka. This determines also the northeasterly course of the eastern coast line.

The third line is that of the island of Sagalin (Krafto), which, trending due north and south, would seem to determine the N. S. course of the western coast line of Yesso, and the N. S. trend of Nippon from its northern point to the Bay of Yedo.

I have already referred the N. E. line of uplift to the Sinian system of elevation, in a previous chapter; the N. W. trend affecting, as it does, the oldest metamorphic rocks, is perhaps older, and the N. S. trend younger.

Neighborhood of Nagasaki, on the West Coast of the Island of Kiusiu.

This port is at the head of a long narrow inlet, or *fiord*, which has nearly a N. E., S. W. trend, and lies between long ridges, the peaks of which rise to between 1,000 and 2,000 feet above the sea. The skeleton rocks of these hills are metamorphic strata. These were mica schist dipping vertically, in both the ridges where they were examined, northwest and southeast from the city, and argillaceous and talco-argillaceous schists, with some limestone, where the eastern ridge was seen near its southern end, opposite the island of Kabasima. On this island the trend of the strata is nearly N., S., and they are traversed by a broad belt of granite bearing fragments of the schists near the planes of contact. On the island Amaksa, a few miles further east, crystalline, white limestone, and a fine sandstone are quarried.

The greater part of the country, in the neighborhood of Nagasaki, is covered, to the summits of the highest hills, with an extensive pluto-neptunian deposit, resembling in general character the volcanic tufa-conglomerate of Yesso.

In places along the eastern side of the bay, and on the islands at its mouth, the rocks of a coal-bearing formation are exposed. Of these only a coarse, hard sandstone, with threads of coal was seen, as it was not permitted to foreigners to land at any of these localities. The position of these beds, however, is such as to make it probable, that the rocks of this coal basin rest immediately, and nonconformably, on the metamorphic strata before mentioned.

In the terraces which in places fringe this coast, we have again evidence of oscillations in level, since the beginning of the volcanic epoch. The terraces are very tufaceous, and seem to be of more recent deposition than the conglomerate that covers the higher hills.

Bay of Yedo.

Nearly all the country included within the treaty limits, or radius of twenty-five miles from Yokohama, which area alone is accessible to foreigners, is of recent formation. A bluff, from 60 to 100 feet high, of bluish clay containing recent shells, and fragments of pumice, with an upper stratum of more gravelly character, faces the bay. From the summit of this bluff a plain of the same deposit extends westward, about twenty miles, rising gently, till the mountains of Oyama. I was not permitted to ascend these mountains, but from the gravels of the streams descending from them I judged them to be metamorphic. The fragments seen were of diorite, gabbro, and serpentine.

South of Yokohama the ridge of the peninsula of Sagami also furnishes fragments of serpentine. The western side of the peninsula, as well as the island of Enosima, are of a firm, fine-grained gray sandstone and conglomerate, in apparently horizontal strata.

Previous to the elevation of the recent beds, the peninsula of Sagami, and probably also the highland east of the Bay of Yedo, were islands.

The existence of these recent marine terraces along the Japanese coast, from Yesso to Kiusiu, and of similar deposits on the China coast, as at Chifu and along the western edge of the great delta plain, point to widely extended changes, in recent times, in the relative position of land and water. A careful study of their characters, as regards the organisms they contain—a study that should include the recent deposits of the Amur system,[1] and perhaps also those of the Manchurian rivers—would probably throw much light on the age of the Gobi desert deposits, and through this on some of the most important questions of quaternary and younger tertiary geology.

---

[1] M. Schmidt observed, almost everywhere on the Amur, between Strelka and Blahowestschensk, terraces of fresh-water tertiary rising nearly 200 feet above the river.—*Peterman's Mittheilungen*, 1861, p. 315.

CHAPTER X.

MINERAL PRODUCTIONS OF CHINA.

THE following list of minerals, and their localities, is compiled from Chinese geographical works, the Tatsingitungchi having furnished the greater part, though for the sake of completeness, the special geographies of the different provinces, and often those of departments, were searched.

The compilation involved the examination, by the author's Chinese secretary, of over one thousand volumes.

Only a portion of the list compiled can be made available for publication owing to our inability to identify the Chinese names for a large proportion of the useful minerals.

The orthography adopted by Dr. S. W. Williams, for Chinese geographical names, is followed in the list, where the subdivision of the country into provinces, departments (Fu), and districts (Chau, Hien, or Ting), is also observed.

List of Localities of Useful Minerals in China.[1]

IRON.

PROVINCE OF CHIHLI.

SHUNTIEN (Fu) or PEKING. At TSUNHWA (chau) WANGPING (hien) at Chingshui near Chaitang. At Tiekung Mt. 30 li E. of MIYUN (hien).

PAUTING (Fu). In MWANCHING (hien).

SIUENHWA (Fu). In LUNGMUN (hien) lodestone.

YUNGPING (Fu) At Mang Mt. 15 li N. E. of TSIENNGAN (hien). At Mt. Tsz' 15 li W. of Lulung (hien), with gold and silver ores.

SHUNTEH (Fu). At Mt. Hai 40 li W. of SHAHO (hien.)

KWANGPING (Fu). Lodestone at TSZ' (chau).

PROVINCE OF SHANSI.

TAIYUEN (Fu). In TAIYUEN (hien) and YUTSE (hien).

PINGYANG (Fu). In KIUHYU (hien). YUTSUNG (hien). YOYANG (hien). KIH (chau). HIANGNING (hien)

PUCHAU (Fu) Hien not indicated.

KIAI (chau). In NGANI (hien).

KIANG (chau) At Mt Kiang 20 W. of KIANG (hien).

LUNGAN (Fu) Hien not indicated.

FANCHAU (Fu). At Siyen Mt in HIAUNI (hien).

---

[1] Localities producing coal, lime, alum, salt, and gold, are tabulated on pages 56–61.

TSEHCHAU (Fu). In YANGCHING (hien).
TATUNG (Fu). In HWAITSUNG (hien).
PINGTING (chau). Hien not indicated.

PROVINCE OF SHENSI.

SINGAN (Fu). Hien not indicated.
SHANG (chau). 180 li N. E. of the city at Mt. Tiling.
PIN (chau). Hien not indicated.
FUNGTSIANG (Fu). In LUNG (chau) and MEI (hien).
HANCHUNG (Fu). In TSUNGKU (hien). At Lotsung Mt. N. W. of SIAYANG (hien). At Tie Mt. 5 li N. of MIEN (hien).
FU (chau). In CHUNGPU (hien) and IKIUN (hien).

PROVINCE OF KANSUH.

PINGLIANG (Fu). In PINGLIANG (hien) and HWATING (hien).
KUNGCHANG (Fu). At Te'yang Mt. 120 S. of NINGYUEN (hien). At Ningkwei Mt. 30 li S. of Ningyuen (hien), with silver and copper ores.
TSIN (chau). In TSINGNGAN (hien) and HWUI (hien).
KINGYANG (Fu). At Mt. Hungling 18 li N. of NGANHWA (hien).
NINGHIA (Fu). Hien not indicated.

PROVINCE OF SHANTUNG.

TSINAN (Fu). In CHICHUEN (hien). At Mt. Chang 50 li S. E. of SINCHING (hien).
TAINGAN (Fu). In LAIWU (hien); S. E. 13 li at Mt. Tashï, and N. W. 3 li at Mt. Kung.
YENCHAU (Fu). In YIH (hien).
ICHAU (Fu). At Mt. Chipau 100 li N. of KÜ (chau) in vicinity of gold, silver, copper, lead, and tin ores.
TSINGCHAU (Fu). A Mt. Tie 90 li from YIHTE (hien). In KAUYUEN (hien) and LONGAN (hien). At Mt. Chang in LINGTSE (hien). At Mt. Sung 60 li S.W. of LINKÜ (hien) in the vicinity of silver, lead, copper, tin, and cinnabar ores and gold washings.
TUNGCHAU (Fu). In PUNGLAI (hien).

PROVINCE OF KIANGSUH.

KIANGNING (Fu) or NANKING. At Tsz Mt. in KIUYUNG (hien), with copper ores. Lodestone at Mt. Yen in LUHHOH (hien).
CHINKIANG (Fu). 30 li S. W. of LIYANG (hien).
HWAINGAN (Fu). In YENCHING (hien).
SÜCHAU (Fu). At Mt. Pema 90 li N. E. of TUNGSAN (hien).

PROVINCE OF NGANHWUI.

NGANKING (Fu). Hien not indicated.
TAIPING (Fu). Steel works at Tekang in FANCHANG (hien).

PROVINCE OF HONAN.

HONAN (Fu). In the hiens, KUNG, NIYANG, TUNGFUNG, SINGAN, and SUNG
NANYANG (Fu). In the hiens, NANYANG and NEYANG.
KAIFUNG (Fu). In YU (chau).
CHANGTEH (Fu). In SHEH (hien).
JU (chau). Hien not given.

PROVINCE OF HUPEH.

WUCHANG (Fu). In KIANGHIA (hien) and WUCHANG (hien). At Mt. Hwuilu E. of TAYÉ (hien). At Mt. Tsz'hu 50 li N. E of TAYÉ (hien) lodestone. At Hwangko Mt. 2 li W. of HINGKWOH (chau), in vicinity of silver ores.

HWANGCHAU (Fu). At Mt. Kung 40 li W. of MACHING (hien). At Mt. Kung 15 li S. E. of HWANGMEI (hien).

PROVINCE OF SZ'CHUEN.

CHINGTU (Fu). In TSINGTSING (hien).

TSZ' (chau). Hien not indicated.

MIEN (chau). Hien not indicated.

NINGYUEN (Fu). In HWUILI (chau), MIENNING (hien), and YENYUEN (hien).

PAUNING (Fu). In KWANGYUEN (hien).

SHINGKING (Fu). Hien not indicated.

CHUNGKING (Fu). At Mt. Tie 80 li S. E. of YUNGTSANG (hien). In HOH (chau). In TUNGLIANG (hien).

CHUNG (chau). In FUNGTU (hien).

KWEICHAU (Fu). In WUSHAN (hien) and YUNYANG (hien).

SUITING (Fu). In KÜ (hien) and in TATSOH (hien).

LUNGNGAN (Fu). Hien not given.

TUNGCHUEN (Fu). In YENTING (hien) and SHÏHUNG (hien).

KIATING (Fu). 40 li N. of WEIYUEN (hien). 100 li N. of YÚNG (hien).

KUNGCHAU (Fu). At Kusung Mt. 10 li S. of the city in vicinity of copper ore.

PROVINCE OF KIANGSI.

NANCHANG (Fu). In FUNGSIN (hien) and TSINHIEN (hien).

KWANGSIN (Fu). In YOHYANG (hien), YÜSHAN (hien), KWEICHI (hien), and SHANGTSAO (hien).

KANCHAU (Fu). At Tishan in WEITSANG (hien).

NANNGAN (Fu). In TAYÜ (hien).

PROVINCE OF HUNAN.

CHANGSHA (Fu). Hien not given.

SHINCHAU (Fu). Hien not given.

HANGCHAU (Fu). Hien not given.

YUNGCHAU (Fu). Hien not given.

YUNGSHUN (Fu). Hien not given.

PAUKING (Fu). Hien not given

CHANGTEH (Fu). Hien not given.

CHIN (chau). Hien not given.

TSING (chau). Hien not given.

LI (chau). Hien not given.

KWEIYANG (chau). Hien not given.

YOCHAU (Fu). Hien not given.

PROVINCE OF KWEICHAU.

SZ'CHAU (Fu). At Mt. Lungtang E. of the city, in vicinity of lead ores.

TUNGJIN (Fu). 100 li W. on Sungchi river, in vicinity of gold washings. 140 li W. in the Tichi river.

LIPING (Fu). Hien not indicated.

SHIHTSIEN (Fu). Hien not indicated.

TATING (Fu). In WEINING (chau).

SZ'NAN (Fu). In NGANHWA (hien).

PROVINCE OF CHEHKIANG.

KIAHING (Fu) In HAIYEN (hien).
TAICHAU (Fu). At Lungsu Mt. in NINGHAI (hien), in vicinity of copper ore.
YENCHAU (Fu). At Mt. Tie in KIENTE (hien).
WANCHAU (Fu). In PINGYANG (hien) In TISUNG (hien). In SUINGAN (hien).
CHUCHAU (Fu). In SIENPING (hien).

PROVINCE OF FUHKIEN.

FUHCHAU (Fu). In the hien FUHTSING and MING.
TSIENCHAU (Fu). In the hien TUNGNGAN and NGANCHI.
KIENNING (Fu). In the hien KIENNGAN, TSUNGHO, WUNING, and SUNGCHI.
YENPING (Fu). In the hien NANPING, YUKI, and TSIANGLOH.
TINGCHAU (Fu). In the hien HIANGHANG, NINGHWA, and TSANGTING.
CHANGCHAU (Fu). In LUNGCHI (hien)
FUNING (Fu) In NINGTEH (hien).
YUNGCHUN (chau). In TEHHWA (hien).

PROVINCE OF KWANGTUNG.

LIEN (chau). In YANGSHAN (hien).
SHAUCHAU (Fu). In UNGYUEN (hien).
SHAUKING (Fu) In hien YANGTSUNG, YANGKIANG, and SIUHING.
KIUNGCHAU (Fu). Lodestone, locality not given.
LOTING (chau). Excellent ore at Mt. Wutungtu in TUNGNGAN (hien).

PROVINCE OF KWANGSI.

LIUCHAU (Fu). In YUNG (hien).
PINGLOH (Fu). At Chingkang Mt. 120 li S E. of Ho (hien). At Mt. Chaukang 45 li N. E. of Ho (hien).

PROVINCE OF YUNNAN.

YUNNAN (Fu). In KWUNGMING (hien) and YUNGMEN (hien).
LINGAN (Fu). In SINGO (hien) at Hungtonientsa, Sanhotsa, Liulungtsa, and Tsingtsa. In SHIH-PING (chau).
TSUHIUNG (Fu). At TSUYUTSUNG in TINGYUEN (hien). 50 li W. of TSUNGNAN (chau).
CHINKIANG (Fu). In SINGHIUNG (chau).
KIUHTSING (Fu). At Tseh Mt in SIUENWEI (chau) in vicinity of copper ore. In NANYING (hien), and in the chau LOHLIANG, CHENYIH, MALUNG, and NANYING.
WUTING (chau). Iron ore and iron works at Tameti (tsang), Tsetse (tsang), Ineh (tsang), Loti (tsang), and Sanpu (tsang). Also in LUHKIUEN (hien) at Tsiehliu (tsang) and Tsutsu (tsang).
YUNGCHANG (Fu). Iron works at Aying.
TUNGCHUEN (Fu). At Mokwei and Tashuitang.
MUNGHWA (ting) In the mountains west of the city.
YUNGPEH (ting). Locality not indicated.

ORES OF COPPER, SILVER, LEAD, TIN, QUICKSILVER.

PROVINCE OF CHIHLI.

SHUNTIEN (Fu) or PEKING. Silver at Mt. Yinyen 15 li S. of MIYUN (hien). Silver at Sz'ling 100 li N. E. of MIYUN (hien).

YUNGPING (Fu). Silver 130 li N. W. of TSIENGAN (hien). Silver at Mt. Tsu 15 li W. of LULUNG (hien), in vicinity of gold and iron ores. Silver at Mt. Yühwang 90 li N. E. of FUNING (hien). Tin in TSIENNGAN (hien).

PAUTING (Fu). Copper.

SIUENHWA (Fu). Silver in YU (chau).

PROVINCE OF SHANSI.

PINGTING (chau) Copper in YU (hien).

TAI (chau). Blue and green carbonates of copper.

PINGYANG (Fu). Copper at Mt. Kiang 20 li S. W. of KIUHIU (hien).

KIAI (chau). Copper in twelve localities. Silver in NGANI (hien). In PINGLOH (hien) silver in several localities, copper in forty-eight localities, and tin at Mt. Ki 60 li N. E. of the city.

KIANG (chau). In YUENCHU (hien). Lead at Mt. Peh, and copper at Mt. Sanchuen 80 li N. of city. Copper in WUNGHI (hien).

LUNGAN (Fu). Copper in all the hien.

TSIN (chau). Tin in TSINYUEN (hien).

TSEH (chau). Copper and tin in YANGCHING (hien).

TATUNG (Fu). Copper. Malachite at Mt. Shïlieu 5 li E. of the city.

PROVINCE OF SHENSI.

SINGAN (Fu). Silver. Copper at Mt. TSUNGNAN 50 li South of city, in vicinity of jade and iron.

SHANG (chau). Cinnabar. In LOHNAN (hien), malachite at Mt. Yih 60 li E. of city. Silver and tin at Mt. To 90 li S. W.; copper 90 li S. E., and at Sihungnien 50 li S. E. of city.

HANCHUNG (Fu). Quicksilver and cinnabar at Mt. Sz'ni N. W. of LIAYANG (hien).

HINGNGAN (Fu). Blue and green carbonates of copper at Mt. CHINGLIEU 45 li E. of city. Cinnabar and quicksilver at Mt. Shuiyin 140 li N. E. of Sinyang (hien).

PROVINCE OF KANSUH.

PINGLIANG (Fu). Silver and copper in PINLIANG (hien). Silver and copper in HWATING (hien).

KUNGCHANG (Fu). Silver and copper at Mt. Ningkwei 30 li S. of NINGYUEN (hien).

KIAI (chau). Quicksilver. Silver at Yinyu 73 li N. W. of WAN (hien).

TSIN (chau). Silver at Mt. Tayang 50 li N. E. of TSINGNGAN (hien). Copper in TSINGNAN (hien). Silver at Mt. Sungkia 90 li N. E. of LIANGTANG (hien). Silver in TSINGSHUI (hien). In HWUI (hien) lead, and at Mt. Chichi, S. of city, cinnabar.

PROVINCE OF SHANTUNG.

TAINGAN (Fu). Copper at Mt. Yingliang 30 li N. of LAIWU (hien).

YENCHAU (Fu). Tin in YIH (hien). Copper at Mt. Koyeh 15 li S. E. of YIH (hien).

ICHAU (Fu). Lead in ISHUI (hien). Silver in vicinity of gold ores, at Mt. Pau 90 li S. W. of LANSHAN (hien). Silver, lead, copper, and tin, as well as gold and iron, at Mt. Chipau 100 li N. of Kü (chau). In MUNGYING (hien), quicksilver at Mt. Hung 30 li N. of city; and silver at Mt. Leanghien 60 li N. W. of city.

TSINGCHAU (Fu). Silver, lead, copper, tin, quicksilver, as well as iron, and gold-sand, at Mt. Sung 60 li S. W. of LINKÜ (hien).

PROVINCE OF KIANGSUH.

KIANGNING (Fu). Copper at LISHUI (hien). Copper in vicinity of iron at Mt. Tsz in KIUYUNG (hien)

SUCHAU (Fu). Copper at Mt. Tung 80 li N. E. of TUNGSHAN (hien).

PROVINCE OF NGANHWUI.

NGANKING (Fu). Cinnabar in TAIHUSZ'.

HWUICHAU (Fu). Silver and lead.

NINGKWOH (Fu). Copper in all the hien.

PROVINCE OF HONAN.

HONAN (Fu). Lead in SUNG (hien), and tin at Mt. Lupan in the same hien.

NANYANG (Fu). Copper at Mt. Chihli in TSINGPING (hien). Tin in YÜ (chau).

CHANGTEH (Fu). Native copper. Tin in WUNGAN (hien).

JU (chau). Tin.

SHEN (chau). Tin in LUSHI (hien) and in LINGPAU (hien).

PROVINCE OF HUPEH.

WUCHANG (Fu). Silver at Mt. HWANGKO 2 li W. of HINGKWOH (chau) in vicinity of iron. Copper in KIANGHIA (hien). Copper in WUCHANG (hien). Copper at Mt. Peisuh 60 li N. of TAYÉ (hien). Tin at Mt. Sieh 5 li S. of FUNGTSUNG (hien).

NGANLOH (Fu). Malachite in TIENMUN (hien).

YUNYANG (Fu). Tin.

PROVINCE OF SZ'CHUEN.

CHINGTU (Fu). Copper in KIEN (chau), and in KINGTANG (hien).

MIEN (chau). Silver. Tin.

NINGYUEN (Fu). Silver at Mt. Miloh 200 li E. of HWUILI (chau). In HWUILI (chau) copper at Fénshuiling 100 li N. of city, and "white copper" (Petung), probably a complex ore, at Mt. Haichi 120 li S. of city. In the same chau green and blue carbonates of copper. "White copper in MIENNING (hien). Copper at Mt. Nan in SICHANG (hien). Silver at Mt. Kohsowa N. W. of YENYUEN (hien).

CHUNGKING (Fu). Copper. Cinnabar in KIKIANG (hien).

YUYANG (chau). Quicksilver and Cinnabar in PANGSHUI (hien).

KWEICHAU (Fu). Tin.

LUNGNGAN (Fu). Tin and Quicksilver.

TUNGCHUEN (Fu). Green and blue carbonates of copper. Copper at Mt. Komung 30 li N. W. CHUNKIANG (hien), also 24 li W. at Mt. Laiyung S., and at Mt. Tungkwei S. W. of the same hien.

KIATING (Fu). Copper at Mt. Tung 120 li S. W. of HUNGYA (hien).

KUNG (chau). Copper, in vicinity of iron, at Mt. Kusung 10 li S. of city.

LU (chau). Blue and green carbonates of copper.

YACHAU (Fu). Copper at Mt. Tung 30 li N. E. of YUNGKING (hien).

MAU (chau). Cinnabar.

PROVINCE OF KIANGSI.

NANCHANG (Fu). Copper at Mt. Si.

JAUCHAU (Fu). In FÉHING (hien), copper, and at Mt. Ying, silver.

KWANGSIN (Fu). Silver at YOYANG (hien) and YUSHAN (hien). Lead in TSIENSHAN (hien).

KIENCHANG (Fu). Silver in NANTSUNG (hien).

FUCHAU (Fu). Copper in LINGTSE (hien). In KINKI (hien) silver, and 120 li E. at Mt. Tung copper.

LINKIANG (Fu). Silver in SANKAU (hien). Copper in SINYÜ (hien)

KANCHAU (Fu). Copper in CHANGNIN (hien).

NANNGAN (Fu). Lead and tin in TSUNGNI (hien).

PROVINCE OF HUNAN.

Changsha (Fu). Silver, copper, lead, tin, and quicksilver.

Shinchau (Fu). Cinnabar. Quicksilver on Luki river.

Hangchau (Fu). Silver, tin, quicksilver.

Yungchau (Fu). Silver, tin.

Yuenchau (Fu). Cinnabar and quicksilver in Tsz'kiang (hien), Funghwang (ting), Yungsui (ting), and Wukang (chau).

Pauking (Fu). Silver. Cinnabar in Wukang (hien).

Chin (chau). Copper, tin, lead, quicksilver, and cinnabar.

Kweiyang (chau). Silver, copper, lead.

Yochau (Fu). Silver.

PROVINCE OF KWEICHAU.

Kweiyang (Fu). Cinnabar and quicksilver in Kai (chau).

Sz'chau (Fu). Lead, in vicinity of iron, at Mt. Lungtang E. of the city. Cinnabar and quicksilver at the Sz'chi river.

Tungjin (Fu) Cinnabar and quicksilver at Mt. Tawan 3 li S. of city.

Shihtsien (Fu). Cinnabar and quicksilver.

Tating (Fu). Copper in Weining (chau).

Tsuni (Fu). Quicksilver and Cinnabar.

Sz'nan (Fu) Cinnabar at Mt. Nitan 5 li S., at Mt. Ningtsing 30 li N E., and 50 li N. E. of Wuchuen (hien). Quicksilver at Moyu, Pangtsang, and Nientau, in Wuchuen (hien).

Hingi (Fu). Quicksilver in vicinity of realgar, at Mt. Peinien. Cinnabar at Lamotsang.

Tuyun (Fu). Lead at Mt. Hianglu in Chingping (hien).

PROVINCE OF CHEHKIANG.

Kiahing (Fu). Copper at Mt. Tsang in Haiyen (hien).

Huchau (Fu). Copper and tin in Anki (hien). Copper in Wukang (hien) and Changhing (hien).

Ningpo (Fu). Tin, in vicinity of gold, on Mt. Kehyu. Copper in Funghwa (hien).

Shauhing (Fu). Copper at Soyachi. Tin at Mt. Tsoking. Quicksilver at Mt. Lungkien in Yüyau (hien).

Taichau (Fu). Silver and lead at Mt. Tientai and Mt. Tsz'nien in Tientai (hien). Copper, in vicinity of iron, at Mt. Lungsu in Ninghai (hien).

Küchau (Fu). Silver ore, yielding $300 to the ton, at Mt. Yinkung in Changshan (hien). Copper at Mt. Tung in Singan (hien). Silver at Mt. Yinkung in Suingan (hien).

Yenchau (Fu). In Kiente (hien) copper in Mt. Tungkwei; and silver in Mt. Yin.

Wanchau (Fu). In Pingyang (hien) silver at Mt. Chauki, Mt. Tsz'ye, and Tientsingyang. Silver on the Chauchi river in Tisung (hien).

Chuchau (Fu). Copper at Mt. Tung in Lungtsiuen (hien). Tin and lead in Sungyang (hien).

PROVINCE OF FUHKIEN.

Kienning (Fu). Silver in the hien, Kienngan, Kienyang, Pusung, and Tsungho. Copper in Kienyang (hien).

Yenping (Fu). Copper in the hien, Nanping, Sha, and Yuki.

Yungchun (chau). Lead in Tating (hien).

Lungngan (chau). Lead in Santsingming and Tsiweitsz'kung.

Tingchau (Fu). Silver at Lungmuntsang in Ninghwa (hien). Silver at Wangpeitsang and Nganfungtsang in Tsangting (hien). Tin at Hiangpau Mt. in Tsangting (hien).

PROVINCE OF KWANGTUNG.

Kwangchau (Fu) or Canton. Silver at Tashuikung in Nanhai (hien) and at Peyinkung in Sinhwui (hien).

Lienchau (Fu). Silver. Tin at Sangpuhia and Singtanghia in Yangshan (hien); in the same hien lead and cinnabar.

HWUICHAU (Fu). Tin of excellent quality in HOYUEN (hien) and YUNGNGAN (hien).
KIAYING (chau). Tin in SANLO (hien) and HINGNING (hien).
SHAUKING (Fu). Silver at Yinkung in KAUMING (hien).
KIUNGCHAU (Fu). Blue carbonate of copper. Silver at Litien in YAI (chau).

PROVINCE OF KWANGSI.

KWEILIN (Fu). Silver and Cinnabar.
LIUCHAU (Fu). Silver in SIANG (chau).
KINGYUEN (Fu). Silver at Mt. Mongin 35 li N. W. of HOCHI (chau). Tin at Kaufungkung 13 li W. and Singchaukung 2 li W. of HOCHI (chau). Cinnabar at Mt. Hi N. of Ishan (hien), and at Mt. Kusih in SZ'NGAN (hien).
SZ'NGAN (Fu). Lead in SHANGLING (hien).
PINGLOH (Fu). Silver in PINGLOH (hien). Silver and tin in FUCHUEN (hien). Silver at Taiping-yintsang in Ho (hien). Copper at Mt. Kü 35 li N. E. of Ho (hien). Tin at Tungyuyen and at Lungtsungyen N. of Ho (hien).
YUHLIN (chau). Cinnabar and quicksilver at Mt. Tungshi 15 li E. of PEHLIU (hien).
SINCHAU (Fu). Silver and lead in KWEI (hien).

PROVINCE OF YUNNAN.

YUNNAN (Fu). Copper in KWUNGMING (hien) and YUNGMEN (hien). Malachite in LIUTSZ' (hien), WUTING (hien), and LUFUNG (hien).
LINGAN (Fu). Copper and Tin in MUNGTSZ' (hien).
TSUHHIUNG (Fu). Silver in KWANGTUNG (hien), and at Soyangtsang and Malungtsang in NGAN (chau), and with lead at Yuntsungtsang in TSUHHIUNG (hien).
CHINGKIANG (Fu). Copper in Lunan (chau).
KWANGSI (chau). Silver and lead at Mt. Peting. Copper at Mt. Chung. Tin at Mt. Shïpau.
KIUHTSING (Fu). Silver and lead at Mt. Yang W. of SIUENWEI (chau). Copper in PINGI (hien).
WUTING (chau). Silver in Sutsuweitsang Copper at Pauhung and Olo. Lead at Mt. Kauyin.
PU'RH (Fu). Silver, lead, and copper at Pema, Kanku, and Mantau in SIHMA (ting). Copper of best quality at Tsilitutsz'.
YUNGCHANG (Fu). Silver at Mingkwang and Aying. Copper and tin at TANGYUEH (chau).
TUNGCHUEN (Fu). Silver in WEITSZ' (hien). Mines of Petung ("white copper") at Tangtangtsang and Talütsang.
CHAUTUNG (Fu). Silver at Lutientsang and Lomatsang, at Tungputsang in CHINHIUNG (chau), and at Kinshatsang in YÜNSEH (hien). Copper at Changfapu in CHINHIUNG (chau), at Siaunienfang in YÜNSEH (hien), and at Ninglau Mt. and Tsietsz'tang in TAKWAN (ting).
YUNGPEH (ting). Copper.

KINGDOM OF COREA.

Gold, silver, quicksilver, iron, coal, and sulphur.

MISCELLANEOUS MINERALS.

PROVINCE OF CHIHLI.

TAMING (Fu). Nitre on the Siau Ho.
SIUENHWA (Fu). Rock-crystal at Mt. Hwangtsie N. of city. Agates at Sz'kiautungtsing.

PROVINCE OF SHANSI.

TATUNG (Fu). Agates, sulphate of iron.
KIANG (chau). Sulphate of iron.
LUNGAN (Fu). Amber.
FANCHAU (Fu). Gypsum. Nitre. Rock-crystal in YUNGNING (chau).
TSEHCHAU (Fu). Rock-crystal. Realgar.

PROVINCE OF SHENSI.

SINGAN (Fu). Jade, in vicinity of copper and iron, at Tsungnan 50 li S. of city, at Mt. Lantien 30 li E. of LANTIEN (hien), and at Mt. Li, in vicinity of gold 2 li W. of LINGTUNG (hien).

SHANG (chau). Jade, in vicinity of gold, at Mt. Yanghwa N. E. of LOHNGAN (hien).

KIA (chau). Agate in FUKUH (hien) and SHINMUH (hien).

HANCHUNG (Fu). Amber in many localities. Feitsui (jadeite) in LIAYANG (hien). Realgar at Mt. Futu 60 li S. of FUNG (hien).

HINGNGAN (Fu). Jade at Mt. Ching 58 li W. of SINYANG (hien), and at Kantientsuhtung 60 W. of PEHHO (hien).

FU (chau). Iron pyrites and sulphur.

PROVINCE OF KANSUH.

KUNGCHANG (Fu). Agates. Realgar at Mt. Leangkung S. W. of MIN (chau). Nitre in NINGYUEN (hien), and HWUINING (hien).

KIAI (chau). Realgar. Sulphate of iron.

KINGYANG (Fu). Nitre in every Hien. Inkstone slate in NING (chau).

PROVINCE OF SHANTUNG.

TAINGAN (Fu). Amethyst.

YENCHAU (Fu). Amethyst.

ICHAU (Fu). Amethyst.

TUNGCHAU (Fu). Gypsum.

PROVINCE OF HONAN.

Nitre in all parts of the province.

PROVINCE OF HUPEH.

ICHANG (Fu). Agates. Nitre.

PROVINCE OF SZ'CHUEN.

CHUNG (chau). Amber in LIANGSHAN (hien).

KWEICHAU (Fu). Amber in WUSHAN (hien) and in TANING (hien).

SUITING (Fu). Amber in TATSOH (hien) or TA (hien).

MEI (chau). Nitre.

PROVINCE OF KIANGSI.

KWANGSIN (Fu). Rock-crystal in SHANGTSAU (hien).

PROVINCE OF HUNAN.

YUNGSHUN (Fu). Nitre in PAUTSING (hien).

YUENCHAU (Fu). Rock-crystal in YUNGSUI (ting).

PROVINCE OF KWEICHAU.

NGANSHUN (Fu). Amethyst.

HINGI (Fu). Realgar at Mt. Peinien.

TSUNI (Fu). Realgar 20 li E. of TUNGTSZ' (hien).

SZ'NAN (Fu). Jade in YINGKIANG (hien).

PROVINCE OF CHEHKIANG.

HANGCHAU (Fu). Gypsum at Mt. Shïkau in SUNGHO (hien).

KÜCHAU (Fu). Lapis-lazuli at Mt. Nien in CHANGSHAN (hien).

YENCHAU (Fu). Rock-crystal in SUINGAN (hien).

WANCHAU (Fu). Lapis-lazuli on Kinchingshï river, in LOTSING (hien).

PROVINCE OF FUHKIEN.

CHANGCHAU (Fu). Rock-crystal in CHANGPU (hien).
TAIWAN (Fu). Sulphur in CHANGHWA (hien).

PROVINCE OF KWANGTUNG.

KWANGCHAU (Fu). Amber. Amethyst at Mt. Pau in TUNGWEI (hien).
SHAUCHAU (Fu). Sulphate of iron.
KIUNGCHAU (Fu). Flint at Mt. Li. Whetstone at Mt. Shï. Large rock-crystals at Mt. Wutsz'.

PROVINCE OF KWANGSI

SZ'CHING (Fu). Realgar.
WUCHAU (Fu). Rock-crystal W. of TSANGHOH (hien).

PROVINCE OF YUNNAN.

YUNNAN (Fu). Nitre in YUNGMEN (hien).
WUTING (chau). Blue jade in Tungsan. Touchstone in the Kinshakiang river. Nitre, from wells, in YUENMAU (hien).
LIKIANG (Fu). Green and black jade in Mt. Mohpeh.
YUNGCHANG (Fu). Amber in TANGYUEH (chau). Agates at Mt. Manau in PAUSHAN (hien). Topaz and rock-crystal at Mungmitosz' in PAUSHAN (hien). Feitsui, and white and black jade at Maumotosz', and blue jade at TUNGYUEH (ting).

The mountains of Southern Yunnan seem to abound in precious stones.

The working of beautiful stones into objects of ornament, forms an important branch of industry in several of the large cities. Jade of various colors, serpentine, steatite,[1] and dendritic marbles, are made into an endless variety of household ornaments. Topaz, aqua-marine, pink turmaline, opaque sapphires, jadeite[2] (Feitsui), lapis-lazuli, sungurshï, a mineral similar to turquois, rock-crystal, garnets, and many other precious and semi-precious stones, are carved, with great labor and patience, in very intricate forms. Several snuff bottles carved out of blue corundum were seen, the cavity being very small at the neck, and enlarged symmetrically and polished in the interior.

No diamonds were seen in any of the lapidaries' shops, although the Chinese have a name for that stone. Emeralds are very rare, and although the Chinese name is lieupaushï (green precious stone), they are known among lapidaries as Sz'mulu, the name of Sumatra, whence they are probably obtained.

Rubies are more common, although often confounded with spinelles and hyacinths. Sapphires are frequent, and often of fine water and respectable size.

---

[1] Much of the stone known as pagodite has been shown by Prof. G. J. Brush to be a compact pyrophyllite.

[2] Feitsui is, perhaps, the most prized of all stones among the Chinese. The *chalchihuitl*, a precious stone of the ancient Mexicans, as I have seen it in a mask preserved in the museum of Pract. Geol. in London, and in several ornaments in the collection of Mr. Squiers in New York, is, apparently, the same mineral. This fact is the more remarkable, as there is no known occurrence of this mineral in America.

APPENDIX.

APPENDIX No. 1.

Description of Fossil Plants from the Chinese Coal-Bearing Rocks.

By J. S. Newberry, M. D.

Cleveland, Ohio, September 25th, 1865.

Raphael Pumpelly, Esq.

Dear Sir: The fossil plants you were kind enough to submit to me for examination, though few in number and somewhat fragmentary, have proved to be of very special interest, since they supply the necessary data for determining, approximately, the age of the strata from which they were taken; and rather unexpectedly prove a large part of the great coal fields of China to be of Mesozoic age.

This conclusion is based on the entire absence of Carboniferous plants from the collection; and the presence of well-marked Cycads—species of *Podozamites* and *Pterozamites*, closely allied to, if not identical with, some heretofore found in Europe and America.

I give below, such descriptions of the several species contained in the collection, as could be framed from the somewhat meagre material submitted to me. Future observations, made upon a larger number of more perfect specimens, will be necessary before questions of specific identity or difference can be definitively settled—but it is scarcely probable that any facts, or specimens hereafter to be obtained, will require, modification of the view—that the coal basins which you visited are all Mesozoic and not Carboniferous.

We have, of course, no right to assume from the interesting facts your explorations have brought to light, that no Carboniferous coal exists in China, for it may very well happen, that as in our own country, coal seams of economical value, but of different ages, will be found there, at points not greatly removed from each other. But geologists will not fail to be deeply interested in the fact that so large portions of the coal basins of China, including beds of both anthracite and bituminous coal—worked for hundreds of years, probably the oldest coal mines in the world—are wholly excluded from the Carboniferous formation. So large is this coal-bearing area, indeed, that when joined to the Triassic, Cretaceous, and Tertiary coals of North America, they quite overshadow the Carboniferous coals of Europe and the Mississippi valley, and suggest the question, whether the name given to the formation which includes the most important European strata, has not been somewhat hastily chosen.

Another interesting feature in the fossil plants under consideration is the reappearance, at the far distant points from whence they come, of genera so well known in European and American geology—and the entire absence of the species of *Phylotheca*, *Glossopteris*, etc.—which have made the Indian and Australian coal floras so puzzling to the palæontologist. There are fragments of a new generic form—probably a Cycad—in the collection, and some obscure specimens that may represent other plants new to science, but the *Pecopteris*, *Sphenopteris*, *Podozamites*, *Pterozamites*, &c., have a very familiar look; and in their resemblance to well known forms, give fresh evidence of the monotony of the vegetation of the globe, previous to the introduction of the angiospermous forests of the Cretaceous epoch.

Whether the strata which have furnished these plants should be considered Triassic or Jurassic, remains to be determined by future observations, as the fossils as yet obtained can hardly be considered sufficient for the solution of that question.

From the "Kwei basin" we have numerous pinnæ of a species of *Podozamites*, undistinguishable from one found by Prof. Emmons in North Carolina, in strata now generally regarded as *Triassic;*

but associated with these are a few pinnæ of different form—much more elongated and acute—scarcely differing from those of a European Jurassic species (*P. lancolotus*, Lind.), still the evidence of identity is much stronger in regard to the former species than the latter.

From Pyünsz' we have a fine *Pecopteris*, with the falcate pinnules—so characteristic of the Mesozoic species, and indeed very accurately copying the form of *P. Whitbiensis*, a European Jurassic species—but unfortunately the strata which contain this fossil have been much metamorphosed, the coal converted to anthracite, and the nervation of the fern has been entirely obliterated, while the outline remains distinct.

Probably it will be found as difficult, or rather as impossible, in China, as it has been in this country, to identify all the subdivisions of the Mesozoic strata discernible in Europe; yet we shall doubtless gather there new proofs of the constancy of the order of sequence in geological history, and new evidence of the stability of the foundations on which geology, as a science, rests.

I have under my eye, as I write this letter, four collections of fossil plants which, though from very widely separated localities, are curiously linked together. They are :—

1st. Fossil plants, Cycads and Conifers, collected by myself from the gypsum formation (Triassic) at Abiquiu, New Mexico. Of this collection the most conspicuous and interesting plant is *Otozamites, Macombii*, N.

2d. A collection of fossil plants—Cycads and Ferns, received through Prof. Whitney from Sonora, Mexico, where they occur with coal strata and Triassic Mollusks. In this collection *Otozamites, Macombii* is associated with *Strangerites magnifolia*, Rogers, *Pecopteris falcatus*, Emm, and other plants occurring abundantly in North Carolina.

3d. A collection of fossil plants—Cycads, Conifers, and Ferns, from N. Carolina and Virginia, including beside the last two mentioned, and many others which are new, several species, apparently identical with European Triassic plants—of the genera *Haidingera*, *Gutbiera*, *Laccopteris*, &c., and among other Cycads, *Podozamites Emmonsii*, N.

4. The collection made by yourself in China—Cycads and Ferns—in which one of the most distinctly marked plants is *P. Emmonsii.*

In regard to the American localities cited above, there is, perhaps, no good reason for our withholding assent to the conclusion that the rocks furnishing the fossil plants are Triassic, but, when we remember how much difference of opinion there has been, and indeed still is, upon this subject, even in the light of large collections of fossils, we can hardly with propriety offer even a conjecture as to the *precise* age of the Chinese coal strata.

To recapitulate—one species of *Podozamites*, contained in the collection is apparently identical with an American Triassic species; the other more resembles a European Jurassic plant. The *Pterozamites* resembles both Triassic and Jurassic species, but is identical with neither.

The *Pecopteris* has certainly a remarkable likeness to *P. Whitbiensis*, which occurs both in the Liassic and Oolitic floras; and it is not yet certain that it is not also found in the Carolina and Richmond coal basins.

The *Sphenopteris* and *Hymenophyllites* are altogether new, and suggest no affinities of value in this connection, while the *Taxites*, *Equisetites*, &c., are too obscure to afford us any help.

Yours respectfully,
J. S. NEWBERRY

PTEROZAMITES SINENSIS, *Newb.*

PLATE IX, Fig. 3."

Pt. fronde pinnata, parva, pinnis linearibus patentissimis integris, sub-approximatis vel remotis, sæpe curvatis, basi integris, apice rotundatis, nervis distinctis æqualibus simplicibus, rachi longitudinaliter striata.

This is a very neat and well-marked, though miniature species of *Pterozamites*, having the general aspect of *Pt. Oeynhausianus*, Goepp., but being less than half the size of that species, and the pinnæ are not at all decurrent on the rachis.

Perhaps of all known species *Pt. linearis*, of Emmons (Manual of Geol. fig. 194), from the Trias of North Carolina, most resembles this plant; but in that the pinnæ are much more crowded.

In the specimens obtained by Mr. Pumpelly, fragments of a number of different fronds are shown, all of about the same size, so we may conclude that the figure now given is a fair representation of the plant.

Locality.—In brown sandstone, with *Sphenopteris orientalis*, from Sanyü, west of Peking.

PODOZAMITES LANCEOLATUS, *Lindl.* sp.

PLATE IX, Fig. 7.

Zamia lanceolata, LIND. & HUTT. Foss. Flor. Vol. III, fig. 4.
Zamites lanceolatus, MORRIS, An. Nat. Hist. 1841.

I have provisionally, and with doubt, referred a few pinnæ of *Podozamites*, found in the collection, to this species. These pinnæ have almost precisely the form of those figured by Lindley, and are longer and narrower than those of *P. Emmonsii*—being linear-lanceolate, with an acute long drawn point, and an attenuated base.

In one character they differ from both the species to which I have referred; they seem to have been thicker and more coriaceous than either—the nerves being so deeply buried in the parenchyma as to be scarcely visible.

The distinctness of the nerves depends, however, on the surface of the leaflet exposed, and on the manner of fossilization—coarse micaceous shales, like that which contains the impression before us, rarely showing the nervation with distinctness.

The small number of the pinnæ, of the character I have described, in the collection, renders it difficult to determine, with accuracy, their specific relations. Their value, therefore, in a great degree, consists in the evidence they give us of the presence of the genus to which they belong in the rocks from which they were taken.

Locality.—Kwei basin on the Yangtse river, Province of Hupeh, China.

PODOZAMITES EMMONSII, *Newb.*

PLATE IX, Fig. 2.

P. fronde pinnata, pinnis distantibus integris alternis oppositisve, lanceolatis, apice attenuatis acutis, basi cuneatis, nervis crebris.

This is, apparently, the same plant as that described and figured by Prof. Emmons (Geol. N. Car. p. 331, pl. iii, fig. 7), under the name of *P. lanceolatus;* but that name having been appropriated for another species from the Oolite of Europe, it becomes necessary to give it another.

The specimens which are contained in the collection brought by Mr. Pumpelly, consist mostly of detached pinnæ, scattered in confusion over the surface of pieces of blue shale. These pinnæ agree perfectly in form and nervation with those of the Carolina plant. They are lanceolate in outline, and rather abruptly narrowed to an acute termination at either end. The nerves are fine and numerous, but distinctly visible, converging to a common point at the remote extremity. The rachis to which all were, and a few are still attached, was slender, and striated longitudinally. The specimen figured by Prof. Emmons is the basal portion of the frond where the rachis is strongest. Higher up this character, to which he attaches some importance, would be lost. The Carolina plant is abundant in the upper plant beds, where it is associated with several species supposed to be identical with some from the Trias (Keuper) of Europe, such as *Pecopteris Stutgardtensis*, *Laccopteris germinans*, &c.; it is, however, not quite certain that there are not also found there some species which are found in the Jurassic of Europe. More careful study of this flora will be necessary before that question can be settled; but the beds which contain *P. Emmonsii* are now generally supposed to represent the Keuper of Europe, and the evidence which this gives, as to the age of the Chinese rocks containing it, so far as it goes, points to the same date for them.

Locality.—Kwei basin on the Yangtse river, Province of Hupeh, China.

Sphenopteris orientalis, *Newb.*

Plate IX, Figs. 1 and 1 *a*.

S. fronde tripinnata, rachide longitudinater sulcata, pinnis lanceolatis vel linearibus, acutis, pinnulis sessilibus summis lobatis, inferioribus laciniatis, laciniis rotundatis, apice sæpe emarginatis nervis tenuis, in lobis dichotomis.

This species is more largely represented in the collection than any other, and yet all the specimens consist of comparatively small fragments of a frond of considerable size.

In nearly all of these specimens a remarkable inequality is observable between the pinnules of the upper and under side of the rachis of each pinna—the upper ones being shorter, broader, and more upright; the lower ones elongated, narrow, and more oblique to the rachis.

Probably this is a constant character in the plant, as examples of similar diversity of form are not wanting among living ferns; but I have seen instances of distortion not unlike this in ferns imbedded in rocks which had been much disturbed.

In general aspect this species is not dissimilar to some Carboniferous ferns, such as *Sph. Schlotheimi*, *Sph. tridactylites*, &c., but it still more resembles the Oolitic species *Sph. denticulata* and *Sph. hymenophylloides*, and the Triassic species *Sph. dichotoma*, Alth. It is also considerably like a Triassic species not yet described, found near Baltimore, Md. From all these, however, it is apparently distinguished by the dissimilarity of form in the pinnules of the upper and lower side of the pinnæ, and by the shape of the lobes of the pinnules. In the upper pinnules the lobes are spatulate; in the lower, fan-shaped. Some of the lobes are straightly emarginate at the summit, but generally they have the appearance of being rounded and entire.

Locality.—Sanyü Chaitang basin, west of Peking, China.

Pecopteris Whitbiensis? *Brong.*

Plate IX, Fig. 6.

From "Piyünsz', west of Peking," in a coarse shale charged with the bitumen driven off from the associated coal seam—now anthracite—is a fragment including several pinnæ of the frond of a large fern, which bears a marked resemblance to *P. Whitbiensis;* so much so, that if the nervation, which is obliterated in the specimen before us, were found to be similar, I should have no hesitation in referring it to that species, as no Carboniferous ferns exhibit that peculiar falcate outline of the pinnules, so marked in *P. Whitbiensis*, *P. dentata*, Lind. (*P. denticulata*, Brong.), etc.

P. Whitbiensis is in Europe found both in the Lias and Oolite, according to Brongniart, but is regarded as distinctly a Jurassic species. It has been supposed to occur in the Richmond coal basin in this country; but *some* of the specimens thought to represent the plant, have been found by Prof. Heer to have a reticulated nervation, and therefore to be, both specifically and generically, distinct from *P. Whitbiensis.* A careful examination of all the specimens collected in this country, supposed to belong to *P. Whitbiensis*, will be necessary before we can decide whether it has indeed been found in the so-called Triassic strata of America; and unfortunately we must wait till other specimens, and such as are in a better state of preservation, shall be brought from China before we can positively affirm that it occurs in the coal strata of that country.

Locality.—Shale over anthracite coal, at Piyünsz', west of Peking, China.

Hymenophyllites tenellus, *Newb.*

Plate IX, Fig. 5.

H. fronde bipinnata, parva, delicatula; pinnis lineari-lanceolatis, pinnulis laciniatis; laciniis filiformis vel spatulatis acutis; sori subrotundi laciniarum apicibus insidentes.

In the plumbaginous schist brought from "Piyünsz', west of Peking," are numerous fragments of a frond of a species of *Hymenophyllites*, which seems to be undescribed. These fragments are so small that no clear idea can be gained from them of the magnitude or form of the frond; but it was doubtless a delicate fern of small size, the pinnules deeply cut into linear or spatulate lobes, those of the fertile portions of the frond being specially slender, and bearing the sori at the extremity of each lobe. A fruit-bearing fragment visible in one of the specimens before us calls to mind Lindley's *Tymfanophora racemosa*, which is now regarded as the fertile portion of the frond of *Coniopteris Murrayana*.

This fossil also occurs at Sanyü, near Chaitang, with *Sphen. orientalis*, thus linking together, geologically, these two localities.

Taxites spatulatus, *Newb.*

Plate IX, Fig. 4.

T. foliis coriaceis lineari-lanceolatis vel spatulatis, curvatis, apice rotundatis, basi cuneatis, nervo medio valde distincto.

In a yellow sandy schist, from near the Futau mine at Chaitang, with pinnæ of *Podozamites*, are numerous linear or spatulate one-nerved leaves, evidently derived from some coniferous tree, apparently of the family of Taxineæ, though larger than the leaves of any of the known Yews.

By their size, curved outline, cuneate base, and their variable width, these leaves bear some resemblance to some of those which have been referred to the genus *Podocarpus*, but with one exception all the described fossil species have been found in Tertiary rocks. The exception referred to is *Podocarpites acicularis*, Andræ, from the Lias of Steierdorf, in which the leaves are very long and narrow, having more the form of those of a pine.

Podocarpus Taxites, Unger (Flor. Foss. v. Sotzka), has almost precisely the form of some of the leaves before us; but it is very doubtful whether that was really a *Podocarpus*.

Brongniart has enumerated in his Prodromus a *Taxites podocarpoides*, from the Oolite of Stonesfield, but no figure or description of it has yet been given. Possibly that species may have relations with the one under consideration, which would give the latter a value in determining the precise age of the rocks which contain it.

APPENDIX NO. 2.

Analyses of Chinese and Japanese Coals.

Made for R. Pumpelly by Mr. James A. Macdonald, M. A., of the Sheffield Laboratory, Yale College.

In the following analyses each determination is the mean of two closely agreeing ones. For the water determination the coal was pulverized and heated in an air-bath at 110° C. until it gave a constant weight. A portion was then ignited in fragments, in a closed crucible, to determine the "volatile matter." The ash was estimated in the usual manner by incineration.

I. Tatsau mine (43 feet seam) near Chaitang.

Hard anthracite. Decrepitates very slightly, and yields a little HO in a closed tube. Spec. grav. 1.57.

Carbon	89.81
Volatile matter	3.08
Water	2.67
Ash	4.44
	100.00

II. Futau mine. Chaitang (west of Peking).

Bright, bituminous, coking coal, yielding a little HO in the closed tube. Spec. grav. 1.30.

Carbon	85.77
Volatile matter	11.94
Water	0.35
Ash	1.94
	100.00

III. Chingshui (near Chaitang W. of Peking).

Soft, bituminous coal, coking in a tube and giving some HO. Spec. grav. 1.37.

Carbon	81.32
Volatile matter	5.62
Water	0.36
Ash	12.70
	100.00

IV. Teyih mine (near Muntakau W. of Peking).

Soft, crumbling anthracite. Gives some HO in a closed tube. Spec. grav. 1.74.

Carbon	80.75
Volatile matter	5.43
Water	2.42
Ash	11.40
	100.00

V. Tashitung mine (Fangshan S. W. of Peking).

Hard anthracite, coated with some carbonate. Decrepitates and gives off some HO in a closed tube. Spec. grav. 1.84.

Carbon	86.62
Volatile matter	4.64
Water	2.64
Ash	6.10
	100.00

VI. Kwei (first mine above Kwei on the upper Yangtse in Hupeh).

Rather a soft coal. When heated in a closed tube gives off HO, and a slightly bituminous odor, without decrepitating. Spec. grav. 1.44.

Carbon	85.63
Volatile matter	4.10
Water	0.38
Ash	9.89
	100.00

VII. Mine of Siangtung (in Hunan).

Hard, fine-grained anthracite. Gives off HO, and decrepitates violently in a closed tube. Spec. grav. 1.65.

Carbon	96.21
Volatile matter	0.65
Water	1.45
Ash	1.69
	100.00

VIII. Another coal from Siangtung.

Hard anthracite. Gives HO in a closed tube, and decrepitates but slightly. Spec. grav. 1.61.

Carbon	94.59
Volatile matter	1.18
Water	1.65
Ash	2.58
	100.00

IX. Laicha Ho (Southern Hunan).

Hard anthracite. Yields HO and considerable sulphur on heating in a closed tube. Spec. grav. 1.47.

Carbon	88.27
Volatile matter	2.92
Water	0.80
Ash	8.01
	100.00

X. HANGCHAU (Southern Hunan).

Rather soft, bituminous coal, coking in a closed tube. Spec. grav. 1.68.

Carbon	71.80
Volatile matter	15.89
Water	0.65
Ash	11.66
	100.00

XI. Mine near FANGSHAN (S. W. of Peking).

Hard anthracite. Yields HO, and decrepitates in a closed tube. Spec. grav. 1.83.

Carbon	90 02
Volatile matter	2.68
Water	2.20
Ash	5.10
	100.00

XII. Coal from TATUNG in Shansi.

Clear black, moderately hard bituminous coking coal. Decrepitates slightly. Spec. grav. 1.30.

Carbon	65.30
Volatile matter	28.69
Water	1.47
Ash	4.54
	100.00

XIII. Coal from DOUY (island of Sagalien).

Clear black, bituminous coking coal. Spec. grav. 1.31.

Carbon	67.51
Volatile matter	22.98
Water	3.51
Ash	6.00
	100.00

XIV. Coal from IWANAI (island of Yesso).

Clear, smooth, black or brownish coal. Gives off HO, and cokes in a closed tube. Spec. grav. 1.26.

Carbon	60.26
Volatile matter	29.72
Water	2.30
Ash	7.72
	100.00

XV. YINGWO mine (Fangshan S. W. of Peking).

Soft crumbling anthracite. Yields considerable HO in a closed tube. Spec. grav. ?

Carbon	77.58
Volatile matter	3.63
Water	2.50
Ash	16.29
	100.00

APPENDIX No. 3.

Letter from Mr. Arthur Mead Edwards on the Results of an Examination, under the Microscope, of some Japanese Infusorial Earths and other Deposits of China and Mongolia.

NEW YORK, January 14, 1866.

RAPHAEL PUMPELLY, Esq.

Dear Sir: I have, agreeably to your request, made a microscopical examination of the specimens of earths you submitted to me some time since, and have to report thereon as follows :—

They were thirteen in number, and the results of examining each one separately and carefully is recorded below. With regard to the two specimens numbered 6 and 9, in which I have found the siliceous loricæ of Diatomaceæ, I have to regret that the time at my disposal lately has been so short that I have been unable to identify the various species detected therein, much less have I been able to do as I would have wished, that is to say, transmit to you at this time a complete list with descriptions and figures of the supposed new forms.

No. 1. "*Efflorescence from the plains of the Kirnoor, Mongolia.*"

This specimen contains some straight sponge spiculæ and broken crystalline particles of a deep olive-green color; otherwise it consists mostly of fine particles of sand. From the presence of the sponge-spiculæ I judge this deposit to be decidedly of aquatic origin and probably marine; although the form of the spiculæ, as well as I can tell from their generally broken condition, is such that they may have belonged to a fresh-water species of sponge.

No. 2. "*Terrace deposit (loam of lower terrace) Tê Hai, Mongolia.*"

Under the microscope this is very similar to the above, that is to say, it contains many of the green crystalline particles found in No. 1, but no sponge-spiculæ that I have been able to detect.

No. 3. *Efflorescence (with sand), from the flat at the Tê Hai Mongolia.*"

This is also very like the first in appearance, in containing green crystals, but, like the second specimen it contains no sponge-spiculæ, so that in neither of these two last numbers have I found anything that would assist in determining their origin.

No. 4. "*Gobi limestone (steppe deposit in part), Nov.* 28, 1864."

Consists almost entirely of fine white particles of calcareous matter, but shows nothing to indicate the circumstances or conditions under which it was deposited. This was to be expected as the microscope rarely reveals anything peculiar in limestones, their origin being best denoted by the character of the large fossils when these are present.

No. 5. "*Lake lòam, Siwan, N. Chihli,*" is mostly sand, and contains a few of the before mentioned green crystals, but no traces of the remains of organized beings.

No. 6. "*Forming bluff near Nietanai, Yesso.*"

No. 9. "*From bluff near Nietanai, Yesso.*"

These both evidently belong to the same deposit, taken at different depths most likely, as is evident from the remains of organized forms which they contain. They are plainly from a marine tertiary stratum similar in character to that discovered by Prof. Rogers underlying the cities of Richmond and Petersburg in Virginia, and also like that found by Prof. W. P. Blake at Monterey in California. The last mentioned deposit I have at present under examination for the State survey of California, and it has been found by Prof. Whitney, and his coadjutors of the survey, at different points extending some hundreds of miles down the Pacific coast, varying slightly in appearance, color, hardness, or the grouping of the forms contained in it, as it was collected at various localities, but plainly showing that there is one extended deposit covering a great extent of country. In fact the Japan specimens resemble those from California in a very marked degree, and much more so than the Virginian ones, containing almost identically the same species of Diatomaceæ that I have found therein. I am not, at present, prepared to give a list of those species, but the following genera have been identified, all of which, with the exception of the last, are exclusively marine, but the species of that last genus *Cocconeis*, found in this deposit, are decidedly of marine origin also.

Arachnoidiscus.	*Creswellia.*
Auliscus.	*Dictyocha.*
Asterolampra.	*Isthmia.*
Actinoptychus.	*Gephyria.*
Aulacodiscus.	*Grammatophora.*
Stictodiscus.	*Rhabdonema.*
Coscinodiscus.	*Biddulphia.*
Triceratium.	*Cocconeis.*

Doubtless species belonging to other genera will be detected hereafter, when I study these specimens more attentively, when it is my intention to make out a full list of the species I may find and publish it, with descriptions and figures of such as I consider new or undescribed, through the medium of some one of our scientific societies. Meantime I send you herewith a couple of slides of this material, mounted in such a manner that you can judge for yourself of its richness in microscopic forms and their beauty, and in many cases, identity with those found in the Californian stratum, a slide of which accompanies them.

No. 7. "*Terrace deposit (loam) from the valley north of the mountains of Sinpaungan.*"
Contains little but sand with a very few of the green colored crystals above mentioned interspersed through it.

No. 8. "*Terrace deposit (loam) from Siwan, N. Chihli, China.*"
This contains nothing of interest or by means of which its origin can be traced.

No. 10. "*Gobi Sandstone, steppe deposit, Dec. 2, 1864.*"
Consists entirely of clean coarse sandy particles, semi-crystalline in character, and with, or in which the microscope reveals, no traces of organic remains.

No. 11. "*From the beds of volcanic ashes at Isoya, west coast of Yesso, Japan.*"
This specimen was examined in a superficial manner at first, but, besides consisting for the most part of pinkish particles of minute size whose origin could hardly be guessed at, was deemed of very little interest. A closer and more thorough examination, however, with higher power glasses revealed decided traces of organic remains and those of an entirely unlooked for character, that is to say, there were found in it, although only in extremely small numbers, straight sponge spiculæ as well as globular, so-called, "gemmules" from sponges, and at the same time dotted ducts from the woody portion of some exogenous plant. Besides these, strange to say, I found fragments of the siliceous epidermis of three or perhaps four species of Diatomaceæ, decidedly aquatic plants and, in this case, all marine in their habit. The genera represented in these very rare and minute fragments were *Arachnoidiscus, Cyclotella, Isthmia,* and probably *Coscinodiscus.* Besides these the green colored crystals mentioned above, as having been detected in several of the earths examined, were seen in this specimen showing that there exists some connection between these various specimens in their origin.

No. 12. "*Alkaline sand from the shore of Lake Kirnoor, Mongolia.*"
No. 13. "*Sand deposited in the valleys around Lake Bilikanoor, Gobi desert.*"
In neither of these specimens could I find the slightest traces of the remains of organized beings or anything else by means of which I could judge of their origin. Thus, although the results of my examination, conducted in the most careful manner, are in most cases but negative, yet, even therefore they are of interest, and you will be better able to judge than I am of their value. The dis-

covery of another marine stratum consisting of the siliceous epidermis of Diatomaceæ in such an unlooked for locality, is of the greatest interest, and will, it is to be hoped, assist somewhat in deciding the true position of such commonly called "infusorial earths." Its similarity to that found on the Pacific coast of North America, would seem to point to its identity in time with that widely extended stratum, and doubtless the results which we have a right to expect from the very complete survey of the State of California, now being carried on, will shed much light on this point. Prof. Toumey placed the stratum of Virginia much lower than had been done by Prof. Rogers, and the correctness or incorrectness of his views in this respect and as bearing on the Californian and Japan deposits, can only be demonstrated after a careful examination and comparison of the adjacent strata. It is desirable that the layer extending from Petersburg in Virginia almost to Baltimore in Maryland, should be examined by a competent observer, and its characters be carefully determined and noted so that they can be compared with those of the Pacific. I hope, ere long, to be able to contribute something towards that end, but extended suites of specimens will have to be collected before we can hope to arrive at any very definite results. Meantime the discovery of such a stratum in Japan will lead to searches for similar deposits in other parts of the world, and I trust and fully expect with success.

Respectfully yours,

ARTHUR MEAD EDWARDS.

INDEX.

F = *Fu*, departmental city; C = *Chau*, sometimes departmental-, but generally district-city; H = *Hien*, district town; T = *Ting*, and Ts = *Tsang*, smaller towns.

PUBLISHED BY THE SMITHSONIAN INSTITUTION,

WASHINGTON CITY,

AUGUST, 1866.

PLATE 1.

SEE CHAPTER II.

Section along the Yangtse Kiang from the Pacific Coast to Pingshan (hien) in Western Sz'chuen.

The portion of the section lying between the coast and the coal-field of Kwei is based on the observations of the author; the remainder is deduced from the observations of Capt. Blackiston, and from the study of the mineral productions of the province of Sz'chuen.

The horizontal distances are taken from the Admiralty charts of the river between the coast and the Tungting lake; thence to Pingshan (hien), from Blackiston's chart of the Upper Yangtse.

The vertical distances east of the Tungting lake are from the Admiralty surveys; west of the Tungting lake they are merely estimated.

PLATE 2.

See Chapter IV.

Route Map of the Yang Ho District.

This map is intended to show roughly the geological and topographical features of a portion of the boundary between the Great Plateau of Central Asia and the mountains of China.

The survey was made by the author from observations with a dioptric compass, the distances being measured by timing a horse whose gait was well known. The work was plotted in the field on a Mercator basis. The route followed in the mountains, immediately west of Peking, is not indicated; on the rest of the map, from Changkiakau (Kalgan) westward, it is marked by the, generally zigzag, line running through most of the villages. Going westward from Changkiakau (Kalgan) by the northern, and returning by the southern route, the plotting overlapped at Changkiakau by five and a half miles, an excess which represents the final, uncompensated, error of the work.

The positions of Siuenhwa, Tatung, and Tungching, are from the Jesuit astronomical observations; that of Peking is from those of the Russian astronomers.

The section lines of Plate 3 are represented on this map.

PLATE 3.

SEE CHAPTER IV.

Geological Sections in Northern Chihli and Southern Mongolia.

Siuenhwa to Daikha Noor.

Nankau to Daikha Noor.

The heights are merely estimated, excepting that of the edge of the plateau, near Hanoor, which is from the measurements of Messrs. Fuss and v. Bunge.

Unfortunately the capital letters indicating breaks in the course of the section lines were omitted on the map, Plate 2.

Plate 3

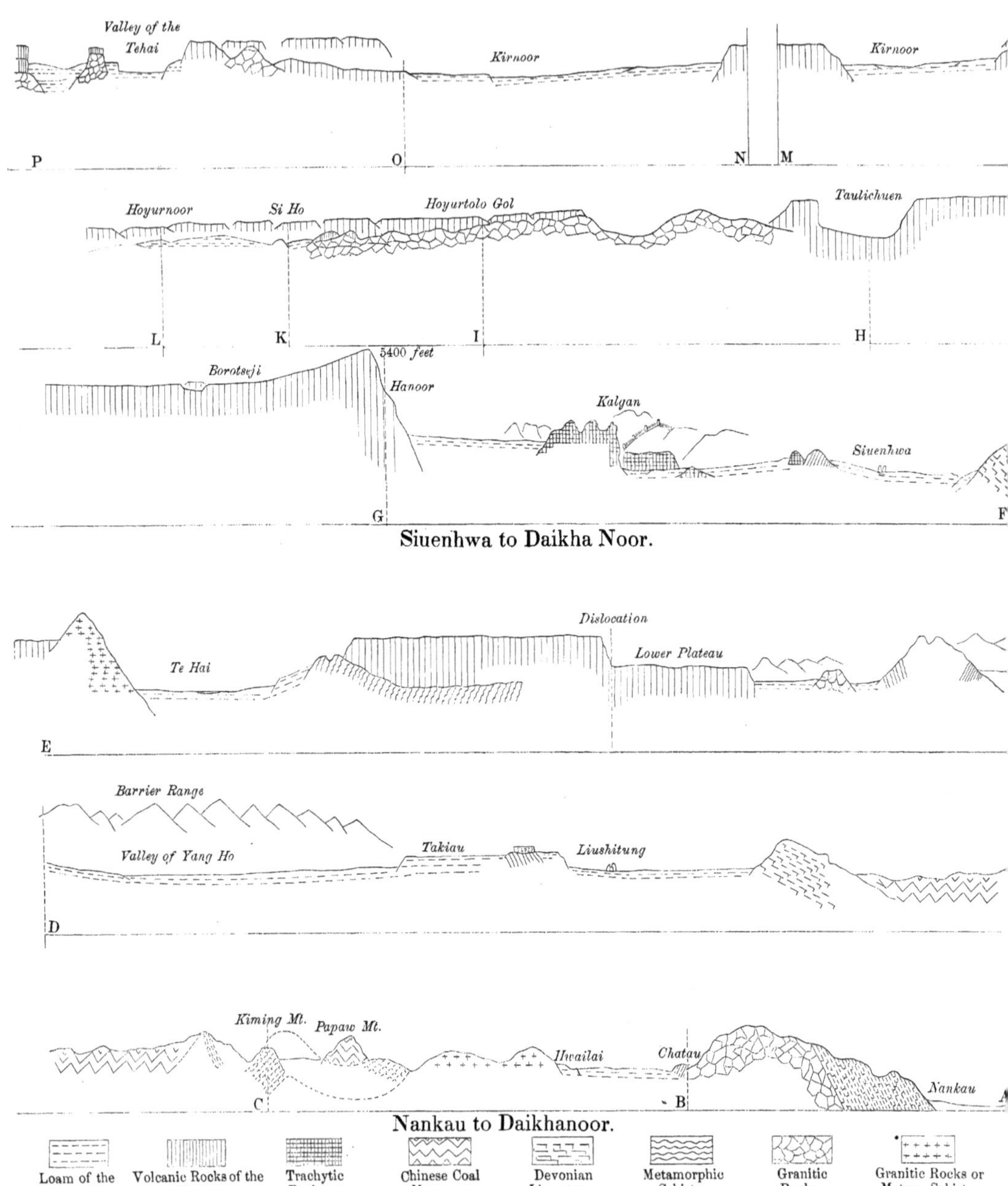

Horiz. Scale 10.46 miles to 1 dec. inch. *Heights* 5000 feet to 1 dec. inch.

PLATE 4.

SEE CHAPTER V.

Maps Representing the Historical Changes in the Course of the Yellow River, or Hwang Ho.

Map I. Lower course of the Yellow river from the time of Yu down to B. C. 602. Also the ancient mouths of the Yangtse Kiang.

Map II. Course after the first great change during the Chow dynasty (B. C. 602).

Map III. Course during the third century, B. C.

Map IV. Course resulting from changes about 132 B. C.

Map V. Second great change about 11 B. C.

Map VI. The channels as they existed during the Tang and five succeeding dynasties, from A. D. 70 to A. D. 1048.

Plate 4

I
II
III
IV
V
VI
Mountainous
PLAIN
DELTA
Kaifung
Tientsin
Hills
Yangtse R.
SHAN-TUNG
Mountainous Region
Gulf of Pechele
Hwang Hai
or
Yellow Sea
30°
35°
118°

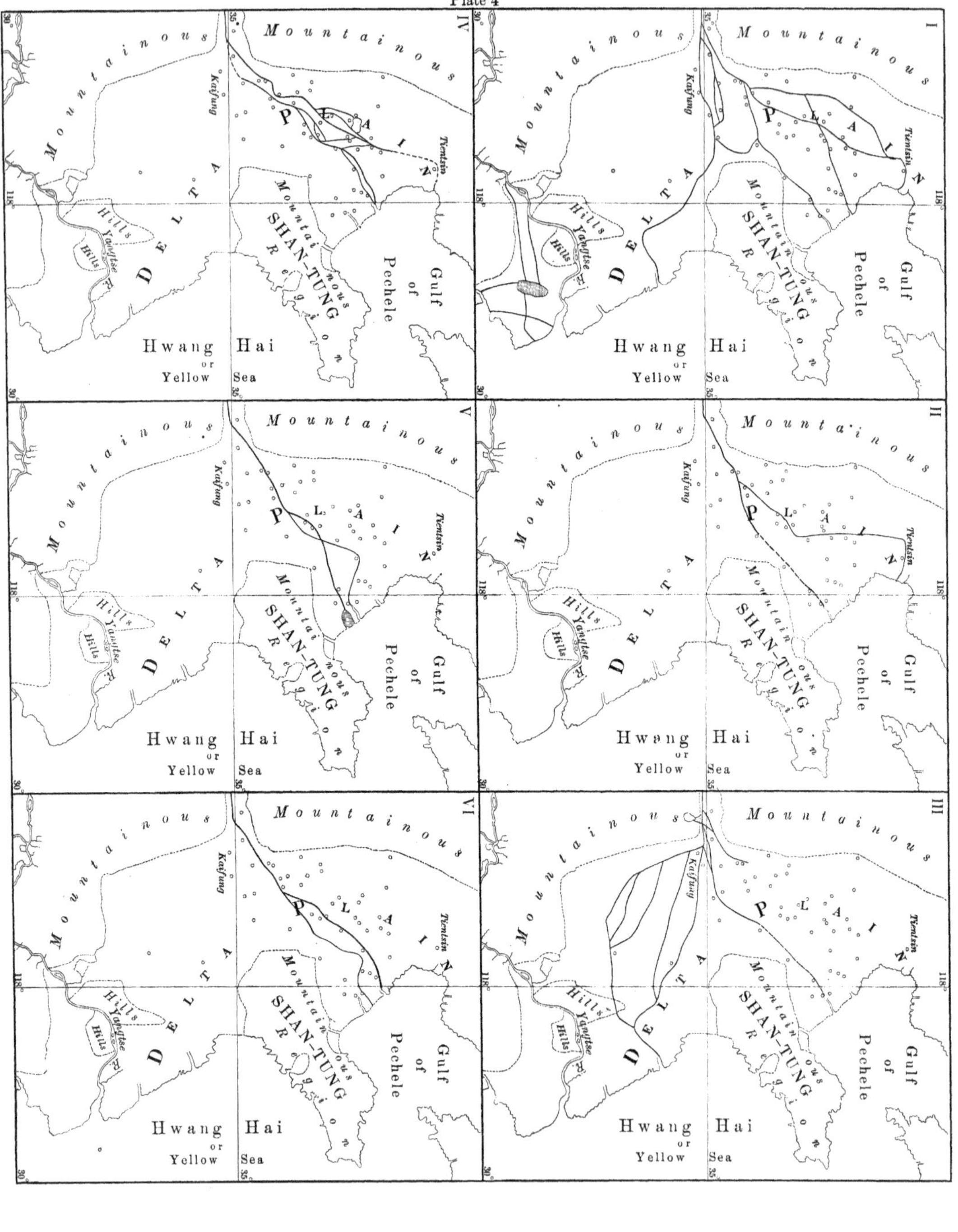

PLATE 5.

See Chapter V.

Maps Representing Historical Changes in the Course of the Yellow River, or Hwang Ho.—Continued.

Map VII. The course under the Sung dynasty, A. D. 1048 to A. D. 1194.

Map VIII. The course under the Kin dynasty.

Map IX. The course under the Yuen (Mongol), and, so far as the channel running due east from Kaifung is concerned, under the Ming and Tatsing (Manchu) dynasties down to the middle of the present century. That portion of the Imperial canal lying north of the Yellow river is indicated, it being mainly in the channel excavated by the river during the Kin dynasty.

Map X. Represents the last change, which occurred within the last ten or fifteen years.

Map XI. Comprehensive map of the Yellow river, including the delta-plain and the ancient lake system, and the supposed former channel of the river through the lakes to the Gulf of Pechele.

Plate 5

VII
VIII
IX
X
XI
Mountainous
PLAIN
DELTA
Tientsin
Kaifung
Gulf
of
Pechele
Mountainous
SHAN-TUNG
Region
Hwang Hai
or
Yellow Sea
Hills
Yangtse R.
Volcanic Plateau of Mongolia
Hwang Ho
Great Wall
SHENSI
SHANSI
Wei R.
Gulf of Pechele
Hwang Hai
or
Yellow Sea
COREA
NIPPON
Ancient Lake System of Northern China.
Supposed outlet of the ancient Lakes before forming of channel between Shensi and Shansi
Delta Plain of the Hwang Ho.

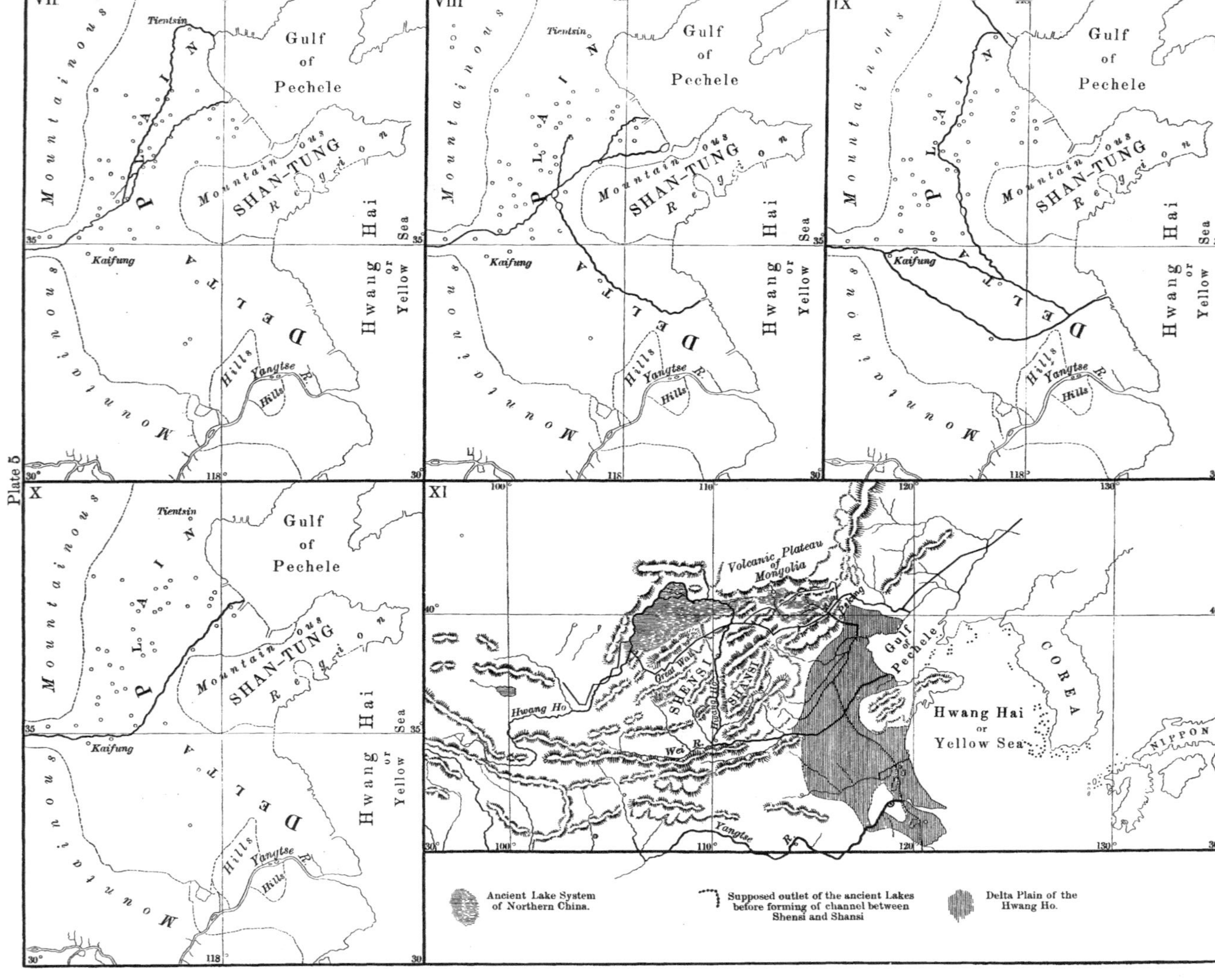

PLATE 6.

SEE CHAPTER VI.

Hypothetical Map of the Geological Structure of China, based on Observations in the North and in the Basin of the Yangtse Kiang, and on a Study of the Mineral Productions of the Empire.

The geographical basis of this map is taken from Arrowsmith's map, published in Blackiston's "Five Months on the Upper Yangtse."

I have altered the position of the Lower Yellow river on the map, to make it agree with its present course.

PLATE 7.

SEE CHAPTER VII.

Map of the Sinian (N. E., S. W.) System of Elevation in Eastern Asia.

The broken line, A, B, indicates the great synclinal axis, and the dotted line, C, D, the main anticlinal axis.

Section across the Table-Land of Central Asia, from the Plain of Peking to near Kiachta in Eastern Siberia.

SEE CHAPTER VIII.

The heights in the northern and southern thirds of the profile are from the measurements of Messrs. Fuss and v. Bunge; those of the central third, being off from their route, are merely approximated.

Plate 7

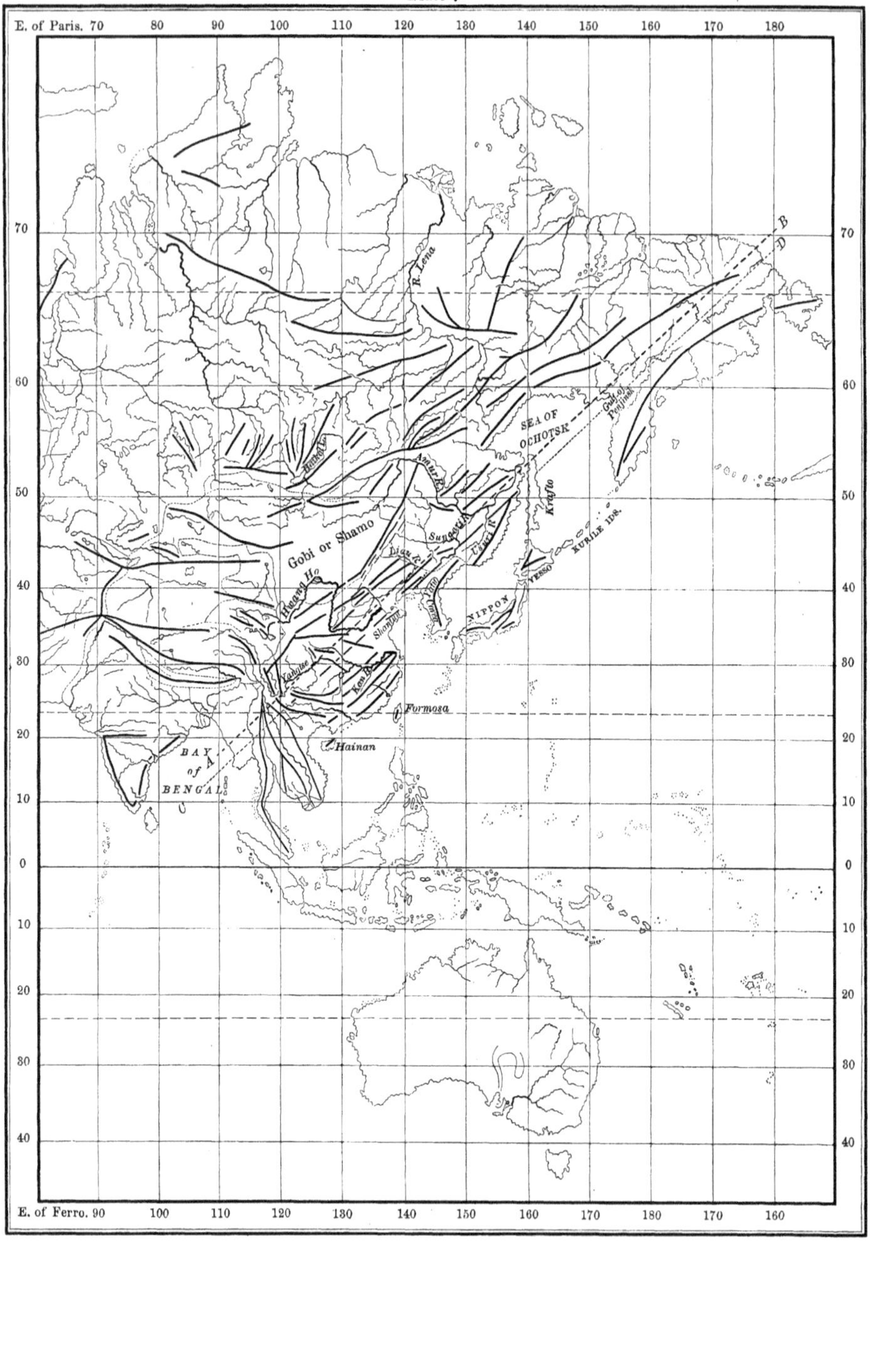

E. of Paris. 70 80 90 100 110 120 130 140 150 160 170 180

R. Lena
Baikal L.
Amur R.
SEA OF OCHOTSK
Gulf of Penjinsk
Kraflo
KURILE IDS.
Gobi or Shamo
Sungari R.
Liau R.
Ussuri R.
YESSO
Hwang Ho
NIPPON
Shantung
Yangtse
Formosa
Hainan
BAY of BENGAL
A
B
D

E. of Ferro. 90 100 110 120 130 140 150 160 170 180 170 160

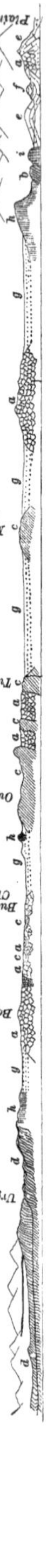

Plain of Peking
Kalgan
Mingan hills
Tugurik
Ouda Urdeni
Budarin-Cheln
Bogdo Ula
Urga

a. Granite. *b*. Metamorphic Schists of N. China. *c*. Metamorphic Rocks of Mongolia. *d*. Metamorphic Rocks of Altai. *e*. Devon Limestone. *f*. Coal Measures.

PLATE 8.

SEE CHAPTER IX.

Geological Route-Sketch. Southern Yesso.

The geographical basis of this map is taken mainly from an unpublished Japanese survey of Yesso, in the Imperial Archives of the vice-royalty of Yesso.

Profile of the West Coast.

Section from the Japan Sea to Volcano Bay.

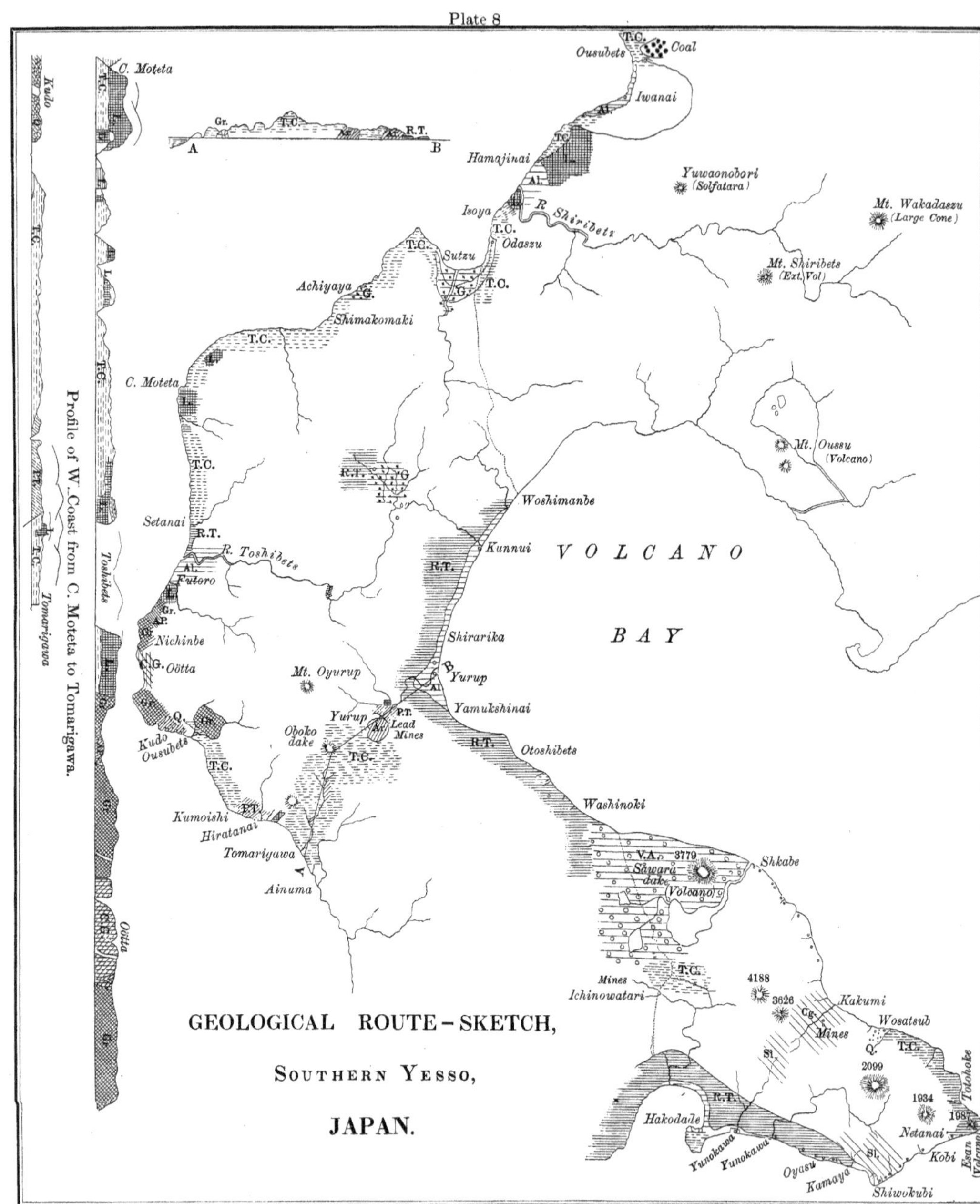

Al. Alluvial and Beach. V.A. *Volcanic Ashes.* G. *River Gravels.* R.T. *Recent Terraces.* L. *Lava.* T.C. *Tufa Conglomerate.* P.T. *Pumice Tufa.* ⁘ *Coal.*
Ar. *Metamorphic Argillite.* Q. *Quartzite.* Sl. Cg. *Clay Slates and Conglomerate.* C.G. *Conglomerate and Granulite*
A.P. *Aphanitic Rock.* Gr. *Granitic and Syenite Series.*

PLATE 9.

SEE APPENDIX No. 1.

Fossil Plants from the Chinese Coal-bearing Rocks.

EXPLANATION OF THE FIGURES.

			PAGE.
Figure	1.	Sphenopteris orientalis	122
"	1*a*.	" "	122
"	2.	Podozamites Emmonsii	121
"	3.	Pterozamites Sinensis	120
"	4.	Taxites spatulatus	123
"	5.	Hymenophyllites tenellus	122
"	6.	Pecopteris Whitbiensis	122
"	7.	Podozamites lanceolatus	121

Plate 9

FOSSIL PLANTS FROM THE CHINESE COAL-BEARING ROCKS.

www.ingramcontent.com/pod-product-compliance
Lightning Source LLC
LaVergne TN
LVHW021358110826
845150LV00007B/1696

* 9 7 8 1 4 2 5 5 1 5 2 0 1 *